Acoustics, Noise and Buildings

Acoustics
Noise and Buildings

P. H. PARKIN
B.Sc., F.I.O.A.

Research Professor,
Institute of Sound and Vibration Research
University of Southampton
and lately Senior Principal Scientific Officer,
Building Research Station,
Department of the Environment

H. R. HUMPHREYS

and

J. R. COWELL
B. Arch., M.Sc., R.I.B.A., M.I.O.A.

Consultant Architect,
Sound Research Laboratories Limited

WITHDRAWN

FABER AND FABER
London · Boston

First published in 1958
by Faber and Faber Limited
Second edition 1961
Third edition 1969
Fourth edition, revised
and reset, 1979
Printed in Great Britain
by Ebenezer Baylis and Son, Limited
The Trinity Press, Worcester, and London
All rights reserved

British Library Cataloguing in Publication Data

Parkin, Peter Hubert
Acoustics, noise and buildings. – 4th ed.
1. Noise control 2. Environmental engineering
(Buildings) 3. Architectural acoustics
I. Title II. Humphreys, Henry Robert
III. Cowell, J R
693.8′34   TH1725

ISBN 0–571–04952–4
ISBN 0–571–04953–2   Pbk

Note

While every care has been taken to ensure that the new edition is accurate and up-to-date throughout, the authors and the publishers can accept no responsibility for any inaccuracies it may be found to contain or for any use or uses to which information or other material contained in the book may be put.

Introduction

The first edition of this book was published in 1958, and in the introduction we said: 'We have tried to give what we believe to be the best practical advice in the light of present knowledge of this difficult but fascinating subject.'; this remains our aim today. Even in 1958 there was more detailed knowledge available than could be conveniently included in one book, so some simplification was unavoidable. This is even more true today when the knowledge has grown so much and the book—if it is to be of maximum practical help—can not be much bigger. Apart from the sheer growth in knowledge, quite a lot of the previous advice needed to be changed in the light of experience and new information (as of course happens in all technical subjects). So we felt that a major revision of the book was called for but—sadly—my original co-author Henry Humphreys died before the revision had progressed far. I was fortunate to be able to persuade Richard Cowell to collaborate with me, and this present, very much revised fourth edition is the result of that collaboration.

The aim of the book, then, remains the same—to provide practical guidance to all concerned with sound and vibration in and around buildings—and we have adjusted the format only so far as current needs demand. New developments are constantly occurring, and legislation and official and unofficial standards change at a rate no text book can keep up with. As a result, in some cases the reader must also consult the latest available literature. Nevertheless, we have tried to cover as clearly as possible the principles and practice at the time of writing.

We believe that clear understanding of the practical aspects is more necessary than ever. We hope that this book contributes to such an understanding.

P. H. PARKIN

Contents

CONTENTS

Partitions—Sliding, Folding Partitions—Flexible Sound-
Insulating Materials—Seals—Acoustic Doors—Acoustic
Screens and Partial Enclosure—Noise Control by
Absorption—Background Sound Control—Enclosures—
Damping—Duct Lagging—Ducts and Pipes leaving Plant
Rooms—Duct Lining and Attenuators—Acoustic
Louvres—Vibration Isolation—Hearing Protection—Noise
from Construction and Demolition

11

CONTENTS

Illustrations

Fig.

Acknowledgements

In a book such as this which covers a fairly wide field, its contents can not be based entirely on the authors' own experiences, and obviously a lot of the material is taken from papers, articles and books published by other people too numerous to mention. The experience of the first author (Parkin) has been obtained from his years at the Building Research Station, and Richard Cowell's experience has been obtained mainly from his employment as company architect to Sound Research Laboratories Ltd., and we are grateful to both these organisations.

We wish to thank *Architectural Design* for Fig. 23; Figs. 67 and 85 to 89 are derived from Crown Copyright material and are reproduced by permission of the Controller of Her Majesty's Stationery Office. We are very grateful to Margaret Humphreys for preparing those drawings which have survived from the original edition, and to Alan Morley for the new drawings; and also to Mary Humphreys and Margaret Morley for typing the text.

1

Nature of Sound

Sound is the sensation produced through the ear resulting from fluctuations in the pressure of the air. These fluctuations can be set up in a number of ways, but usually by some vibrating object, and are in the form of a longitudinal wave motion. Consider the air close to the surface of some object which is vibrating. As the surface moves outwards the air molecules next to the surface are pushed closer together, i.e. the air is compressed. The air cannot move back into its original position for the moment because the space is occupied by the advanced surface of the vibrating object and therefore a movement of air occurs away from the surface. This movement in turn causes the compression of another layer of air. This is followed by a further release of pressure, again by movement of air outwards from the vibrating surface. The result of the outward displacement of the vibrating surface is thus to produce a layer of air compression progressing outwards from and parallel to the surface.

This does not mean that air travels as a whole outwards from the vibrating object. It is only the wave of compression which travels continuously as a result of small limited movements of air molecules in this direction (see Fig. 1). The wave of pressure will move outwards at a steady rate and after it has gone a certain distance, i.e. after a certain time, the surface of the vibrating object will have moved in again through its rest position to one further back. (It is assumed that the surface moves both outwards and inwards from a normal rest position, which is the usual behaviour.) This movement will result in a movement of some air molecules back again (in the reverse direction to the one in which the pressure wave is travelling) and a layer of air of low pressure (or rarefaction) next to the surface. Again, another movement of the molecules of air in the layer next further from the surface back towards the vibrating object will take place. This small movement will also progress outwards (as a chain of events) in the same way as the pressure wave progressed outwards although in this case the movement of the air molecules is in the reverse direction (see Fig. 1). If the vibration of the surface continues

17

its next movement will be outwards, the whole process will be repeated, and a travelling pattern of layers of compression and rarefaction will be established as shown in Fig. 1.

These pressure fluctuations are superimposed on the more or less steady atmospheric pressure and are very much smaller than it. Nevertheless the ear is constructed (as described below) so that it is not sensitive to the steady atmospheric pressure but is able to distinguish these superimposed fluctuations.

The speed of travel of this pattern (in air at atmospheric pressure and temperature of 14° C) is about 340 m/s independent of the rate at which vibrations are being produced. The distance between the

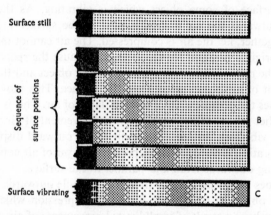

Fig. 1 (A, B, C). Generation and propagation of plane sound waves

layers of compression, i.e. the wavelength, will obviously depend on how fast the vibrations which cause them are taking place. If, for example, one cycle (one complete in-and-out motion) occurs in one-hundredth of a second, then the wavelength will be $\frac{340}{100} = 3.4$ m.

The rate at which the vibrations occur is called the frequency, usually stated in Hertz (cycles per second), and the constant relationship which exists between the quantities is:

$$\text{wavelength} \times \text{frequency} = \text{speed of sound.}$$

The diagrammatic representation of Fig. 1 shows the sound wave travelling in one plane and this is called a 'plane wave'. In practice we are more concerned with what are called 'spherical waves' in which

18

the sound waves travel out from the vibrating object—the source—in every direction. Thus a closer approximation to many practical noise sources would be to replace the plane surface in Fig. 1 by a spherical body which is periodically expanding and contracting. The waves of compression and rarefaction will travel out from the source in a similar manner but they are now spherical and an important practical consequence is that their intensity (i.e. the energy per unit area of the sound wave) will diminish with increasing distance from the source. The reason is that the energy imparted to the air by the source is spread over a greater area as the distance increases, and as the area of a sphere is proportional to the square of the radius it follows that the intensity of the sound waves will decrease inversely with the square of the distance from the source. This is known as the inverse square law and will be referred to later.

When a sound wave meets an obstacle its behaviour will depend on the nature of the obstacle and on its size relative to the wavelength of the sound. This, and the behaviour of sound waves in rooms, is discussed in the following chapters.

Sound sources in practice mostly consist of some vibrating element. For example, in the human voice the air from the lungs is forced past the vocal cords which vibrate and allow an intermittent flow of air to reach the vocal cavities. After modification by these cavities the sound waves are radiated from the mouth. The violin string vibrates when the bow is drawn across it, but the string itself radiates very little sound because it is so small. However, the way in which it is coupled to the body of the violin causes the body to vibrate and this radiates the sound.

Frequency Range

The frequencies we are interested in in this book are those to which the human ear responds and the range is from about 30 cycles per second or hertz (abbreviated Hz) up to about 20,000 Hz, although frequencies higher than 10,000 Hz are seldom important. The corresponding wavelengths are from about 11 m down to about 17 mm. Fig. 2 illustrates this range of frequencies as related to a piano keyboard. The frequency is plotted on a logarithmic scale because the pitch of a sound as heard by the ear varies in this way. For example, every time the frequency of a tone is doubled the pitch goes up one octave. Thus the pitch difference between 100 and 200 Hz is one octave and so is the difference between 1000 and 2000 Hz. On a linear frequency scale the space between 1000 and 2000 Hz would be

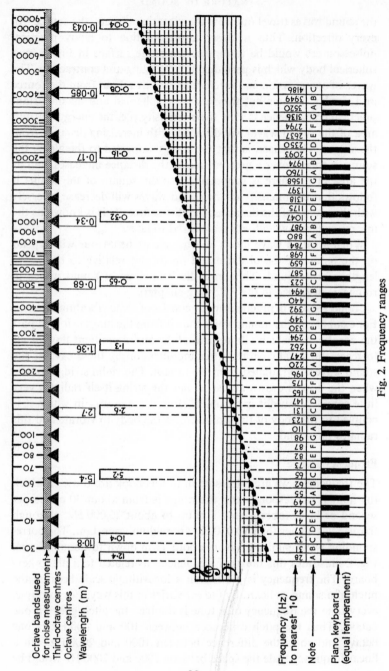

Fig. 2. Frequency ranges

Octave bands used for noise measurement
Third-octave centres
Octave centres
Wavelength (m)

Frequency (Hz) to nearest 1·0
Note
Piano keyboard (equal temperament)

20

ten times that between 100 and 200 Hz whereas on the logarithmic scale the two spaces are equal, to correspond to the pitch change. (Actually, the pitch of a note also depends to a small extent on its loudness, but this need not concern us here.)

The frequencies shown next to the keyboard on Fig. 2 refer to the fundamental (or first harmonic) for each note. However, every musical sound contains in addition to its fundamental a series of overtones (or harmonics) which are multiples of the first harmonic. Thus the second harmonic is twice the frequency of the first harmonic, the third harmonic is three times the frequency, and so on. It is the number and relative strength of the harmonics which distinguishes one musical instrument from another (at least when the instrument is playing a steady note). One of the 'purest' wave-forms, i.e. with little harmonics, is produced by the flute, while the 'richness' of piano tone is due to the large number and comparatively high intensity of its harmonics.

Another difference between musical instruments is their transient response, i.e. the way they behave when first struck or blown. For example, a piano never produces a steady note because the string is either just being struck or is decaying in intensity after being struck. On the other hand, an organ pipe has an initial transient period when it is first sounded but thereafter will continue to emit a steady note.

The Ear

A diagram of the ear is given in Fig. 3, where it is seen that the eardrum is in contact with the air through the auditory canal (called the outer ear) which is about 25 mm long. The atmosphere is pressing on the eardrum with a force very much greater than the force any sound wave will produce, and if there were no compensatory mechanism the eardrum would be pushed hard in. But from the inside of the eardrum (called the middle ear) the Eustachian tube connects with the mouth and allows the atmospheric pressure to operate on the inside of the eardrum also. The result is that the eardrum is not affected by the atmospheric pressure nor by the slow changes which occur in it as the weather changes. The Eustachian tube operates comparatively slowly (and only during the action of swallowing) so that if there is a sudden quick change in the atmospheric pressure, as happens in a rapid descent in an aircraft, then it may not work quickly enough and the ear will hurt. If the Eustachian tube operated instantaneously, i.e. if the inside of the eardrum were connected

21

directly to the atmosphere, then not only would the atmospheric pressure be equalised but the sound pressure would also, and so the ear would not work.

The alternating sound pressures push the eardrum in and out, and this movement, which like all movements and forces involving sound is very small, is transferred by three small bones (the ossicles) joined together (and called the hammer, anvil and stirrup) to a second membrane, the oval window. This second membrane separates the middle ear, which is filled with air via the Eustachian tube, from the inner ear. This mechanical advantage given by these small bones and the

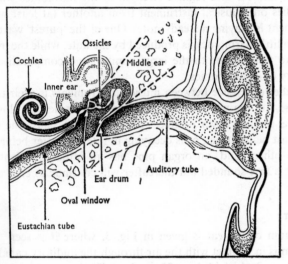

Fig. 3. The ear

fact that the oval window is much smaller than the eardrum combine to increase the sensitivity of the ear.

The oval window is at one end of the cochlea, a hollow member made of bone and filled with liquid. It is spiral-shaped with an unwound total length of about 40 mm. It is divided down the middle by the cochlea partition and part of this partition consists of the basilar membrane on which terminate about 25,000 endings of the main auditory nerve.

Thus an alternating pressure in the air acts first on the eardrum and this in turn acts on the oval window through the action of the ossicles; finally the pressures on the oval window are transmitted to the liquid in the cochlea. The resulting alternating pressures in this liquid are detected by the nerve-endings and transmitted to the brain as the sensation of sound.

Decibels and Phons

The range of sound pressures which the ear responds to is very large although the pressures themselves are small. At this stage, we should explain about the units of sound pressure. Pressures used to be given in dyn/cm^2, but this is now not used, and has been replaced by the pascal (abbreviated Pa) which is 1 $newton/m^2$. 1 pascal is equal to 10 dyn/cm^2, but from now on we will use only pascals.

At the frequencies at which the ear is most sensitive (about 2000 Hz to 4000 Hz) the threshold of audibility occurs at a sound pressure of about 0·00002 to 0·00003 Pa, while the pressure at which the ear begins to hurt is about a million times as great – of the order of 20 to 30 Pa. (For comparison, the atmospheric pressure is about 100,000 Pa.) If sound pressures were given in Pa we would have a very large and awkward range of numbers to deal with. Further, subjectively perceptible increases in loudness are not equal for uniform increases in sound pressure. It is as though the perceived difference in two lengths, one of 20 mm and the other of 30 mm was quite at variance with the perceived difference between another two lengths, one 200 mm and the other 210 mm. In order to obtain roughly equal perceptible differences in sound level the ratio between the two sound pressure quantities must be constant throughout the scale. Thus, to get the same subjective difference between two pairs of sounds, one at 2 and the other at 3 units, the second pair would need to be 20 and 30 units, i.e. so that the second of the two pairs of values has the same ratio to one another (2:3) as the first pair.

This is the basis of the decibel scale which is used to reduce the range of numbers and because it roughly fits the human perception of the loudness of sound.

The illustration (Fig. 4) shows the relation between sound pressures and decibels. This scale is logarithmic (as compared with linear such as a metric rule). For example, a one-decibel increase corresponds to an increase in sound pressure by a factor of about 1·1 (more precisely 1·122).

The decibel is not an absolute measure but always a relative one, i.e. it always gives the ratio of two pressures. For our purposes it will be sufficient to define the decibel as 20 times the logarithm (to the base 10) of the ratio of the two pressures, i.e.

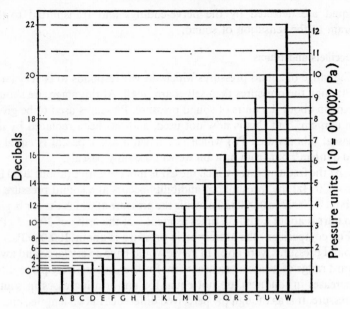

Fig. 4. Relations between sound pressures and decibels

$$\text{decibel ratio} = 20 \log\frac{p_1}{p_2}$$

where p_1 and p_2 are the two pressures being compared. For example, if one pressure is twice the other the decibel ratio is $20 \log (2/1) = +6$ dB. Some commonly occurring ratios are: a factor of $2 = 6$ decibels (abbr. dB), and it follows that a factor of $4 = 2 \times 2 = 6 + 6$ dB $= 12$ dB, a factor of $8 = 18$ dB, and so on; a factor of $10 = 20$ dB and thus a factor of $100 = 10 \times 10 = 20 + 20$ dB $= 40$ dB, and so on up to, for example, the pressure range we have mentioned of $1,000,000$, which equals 120 dB. The larger pressure of the two being compared is so many dB greater than the other, e.g. a pressure of 2 Pa is 6 dB greater than a pressure of 1 Pa. It can also be put the other way round, i.e. a pressure of 1 Pa $=$ a pressure of 2 Pa minus 6 dB.

Sometimes it is only the relative pressures that are required. For example, if sound insulation is involved the pressure on one side of a wall might be, say, 20 Pa (at a particular frequency) and on the other side this might be reduced by the wall to 0.2 Pa. The ratio of the two pressures is 100, or 40 dB. If the first pressure were to be doubled the pressure on the far side of the wall would also go up double and

24

the ratio would still be the same. Thus the pressure reduction due to the wall, or in other words its sound insulation, is always 40 dB no matter what the absolute values of the two pressures are.

However, it is often the absolute value of a pressure that is required and then it is necessary to have a reference pressure if dB's are to be used. There is an agreed reference pressure and it is 0·00002 Pa (i.e. about the threshold of hearing at the ear's most sensitive frequencies). Thus if the noise of, say, a pneumatic drill is measured and found to be 2 Pa then this is a factor of 100,000 greater than the reference pressure, i.e. it is 100 dB greater. The noise is then described as being at a level of 100 dB and when noise levels are given as being so many dB this always means so many dB greater than the reference pressure.

The ear is less sensitive at the lower and higher frequencies (i.e. it takes a greater alternating pressure to produce the same sensation of loudness) than it is at the medium frequencies. However, this difference in sensitivity also varies with how loud the sounds are: only at high sound pressures is the ear roughly equally sensitive to all frequencies.

At 1000 Hz the ear can detect a sound pressure of about 0·00003 Pa (i.e. about 4 dB above the reference pressure). At 50 Hz a greater pressure is required before the ear can detect it—about 0·002 Pa (40 dB above the reference pressure). At 3000 Hz to 4000 Hz the ear is a little more sensitive than it is at 1 kHz and will detect a pressure of about 0·000015 Pa (−3 dB). At the higher frequencies its sensitivity drops, for example, at 10,000 Hz it can detect a pressure of about 0·00015 Pa (17 dB). Of course individuals vary considerably in their sensitivities, and the figures quoted here are typical values.

This threshold of audibility is slightly above the 0 phons contour in Fig. 5. Also shown is the upper limit of hearing, i.e. the levels where the noise is so loud that it begins to hurt—'the threshold of pain'. Between these two limits occur all the pressures we are now concerned with.

We must now deal with how loud a noise of a given frequency and of a given pressure will sound. We have seen that the threshold of audibility is different at different frequencies and similarly the loudness of a sound depends on both its pressure and its frequency. (For the moment we are considering only pure tones, i.e. of one frequency only; these are unlike most sounds and noises occurring naturally or by simple human agency, which are made up of many frequencies.) To form a basis of comparison pure tone of 1000 Hz is used as a

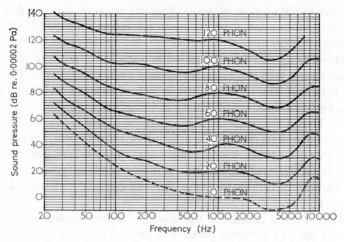

Fig. 5. Equal loudness

reference and a pressure of 0·00002 Pa (0 dB) is the zero on a scale of units called phons. At 1000 Hz the phon level is the same as the decibel level, e.g. a sound pressure of 20 Pa, i.e. 120 dB has a loudness of 120 phons.

At other frequencies than 1000 Hz the ear's sensitivity is different and its judgement of loudness at these frequencies is compared with its findings at 1000 Hz. This comparison must of course be done subjectively and it is usual to use a large number of people who are known to have no defect in their hearing, and to take the average of all their results as being typical of the 'normal' person. This comparison has been done in great detail by many laboratories at various times and although minor discrepancies, some of them unexplained, have appeared the general agreement between results is good.

We have defined the loudness of a tone at 1000 Hz and at a pressure of say 60 dB as being 60 phons, so it follows that if some other frequency tone is adjusted to sound as loud as the 60 phon 1000 Hz tone its loudness must also be 60 phons. Thus a 100 Hz tone at a pressure of 66 dB is found to have a loudness of 60 phons, and a 3000 Hz tone at a pressure of 53 dB also has a loudness of 60 phons.

This sort of comparison has been made at all frequencies and over the whole pressure range the ear is sensitive to, and the results are usually given as in Fig. 5 (derived from the data of Robinson and Dadson). The vertical ordinate shows the pressure levels (in dB greater than 0.00002 Pa); the horizontal ordinate shows the frequency (in Hz): plotted against these two variables are contours of equal loudness. The bottom contour is 0 phons and the other contours go up in 20 phon steps to the top one of 120 phons.

From this graph any single tone can be transposed from dB into phons, or vice versa. For example, if we have a tone at 100 Hz at a pressure of 44 dB, then we can fix this point with reference to the two ordinates and this tone will have a loudness of 30 phons. Similarly, we could say that a tone of 50 Hz at a loudness of 80 phons must be at a pressure level of 94 dB.

It has been mentioned earlier that the difference in the sensitivity of the ear to different frequencies is greatest at low pressure levels, and this is seen in Fig. 5. At 50 Hz a pressure of 46 dB is required to produce a loudness of 10 phons, as compared with 10 dB at 1000 Hz, i.e. a difference of 36 dB for the same loudness; at 50 Hz a pressure of 110 dB is required to produce a loudness of 100 phons, as compared with 100 dB at 1000 Hz, a difference of 10 dB.

Arising out of this it should be noted that a change of pressure level at the low or at the high frequencies produces a greater change in phons, i.e. in loudness, than the same change in pressure level would do at the mid-frequencies. Thus reducing a pressure level of 80 dB at 100 Hz to 70 dB reduces the loudness from about 75 phons to 63 phons, while the same change – 80 to 70 dB—at 1000 Hz reduces the loudness by 10 phons.

The smallest change in loudness of sounds of moderate loudness that the ear can detect is of the order of 0·5 dB. This is so only when the loudness is changed instantaneously or with only a short time interval between changes and it certainly does not mean that such small changes could be detected in practical cases, e.g. when an intruding noise has been reduced by improving the sound insulation.

When we come to deal with noises other than pure tones, i.e. with practically all noises that occur in life, their loudness in phons can still—in theory—be obtained by comparing them with the loudness of a 1000 Hz pure tone, but this is obviously impracticable except under laboratory conditions and then only with noises which do not vary appreciably with time. So in practice, various methods have been developed in the past to calculate the loudness of a noise, in phons, by analysing the frequency content of the noise and by then combining the analysis in such a way as to get a good approximation of the loudness. However, such methods were fairly complicated and it was later discovered that there were much simpler methods with adequate accuracy. The most common of these is to use a sound

27

level meter (described in detail later in Chapter 9) to give the level of a noise in units known as dBA (see Chapters 7 and 9): this figure can be converted into phons quite simply. But later still it became clear that there was not much point in bothering to convert dBA into phons, so the answers are nowadays left in dBA (or the similar units described later) and the levels of various noises in dBA are compared with various criteria, also in dBA.

Masking

Masking is concerned with the effect of one noise on another. For example, while speech will be perfectly intelligible in quiet surroundings it will become less intelligible as the surroundings get noisier until it will be completely obliterated or 'masked' by the noise. Masking is a complicated phenomenon. For example: a pure tone is masked more effectively by another pure tone of nearly the same frequency than by another tone of quite different frequency, and low-frequency noises mask high-frequency noises more than high-frequency noises mask low-frequency noises. In this book we can not go into greater detail about masking, but some practical recommendations are made in Chapter 10.

HEARING

The structure and response of the ear have already been briefly described. We give now a short outline of audiometry and of the more common types of hearing defects. It is intended to serve as a rough guide only (for example to those who may be concerned with the hearing loss of factory workers). A reliable assessment of a person's hearing can only be got by a proper clinical examination.

Audiometry

Audiometry is the term used to describe the measurement of hearing loss, and the instrument used is called an audiometer. This instrument produces pure tones at various frequencies and at pressure levels which can be adjusted over a wide range, but often only in 5 dB steps. The subject wears earphones and, as the level of one of the pure tones is raised, is asked when he can just detect it. This is usually done separately for each ear. If he has no hearing defects then the pressures indicated by the audiometer will be in the region of the bottom curve of Fig. 5. A person who has some hearing defect will not be able to hear these tones at these pressures and the tones must be made louder before the deaf person can hear them. The amount by which the levels must be raised above the normal threshold is defined as the 'hearing loss'. For example, a person whose hearing has been affected by exposure to loud noise (see below) may only just be able to hear at 1000 Hz a level of, say, 34 dB compared with the 'normal' 4 dB, i.e. he has a hearing loss of 30 dB at this frequency; at 4000 Hz the level may have to be 47 dB before he can just hear it, compared with the normal −3 dB, i.e. a hearing loss of 50 dB; and so on at all other frequencies. A graph showing a person's hearing loss as a function of frequency is called an audiogram; the audiogram for the above example is shown in Fig. 6.

It is obvious that when an audiometer is being used the noise level in the room is most important. Although the subject will be wearing earphones for the test, noise may still be loud enough to get through the earphones and 'mask' (i.e. make it more difficult to hear) the test

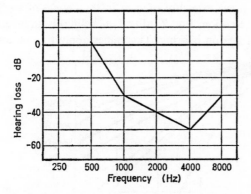

Fig. 6
Example of hearing damage

tones. Thus the test tones may have to be louder before the patient can detect them than they would have to be in quieter conditions; the hearing loss measured under noisy conditions may be higher than the true hearing loss. When general practitioners are using an audiometer they probably make their own adjustment as best they can from their own experience for the local noise conditions. Hospitals, however, will usually have a special room in which the noise level is low enough not to interfere with the measurements.

Another type of audiometric test, not so widely used, is to present recorded speech to the patient via earphones at various loudnesses and to measure the articulation. This is a more complicated procedure than pure-tone audiometry but can provide some indication of the ability of the person to understand correctly words or sentences,

29

i.e. can provide some indication of what practical handicap in understanding speech the person may be suffering from.

Presbyacusis

Presbyacusis is the medical term for the loss of sensitivity of the ear
which occurs naturally with increasing age. There is a complication
when we come to decide what is the 'normal' loss due to age, and it is
that in testing any sample of the population there will be some whose
hearing loss will be due not only to age, but will be due also to, say,
exposure to loud noises at some period in their life which they may
not mention to the tester, or which they may not remember. The
older the age group being tested the higher proportion of such people
there are likely to be and thus the results may show, on the average,
a greater hearing loss than that due to age alone. And, in general,
men show a greater hearing loss with age than women, and this may
be due to a greater noise exposure of men in the course of work,
military service or recreation. Fig. 7 (after Hinchcliffe) is an assessment of the effect of age on hearing. The values apply to men and

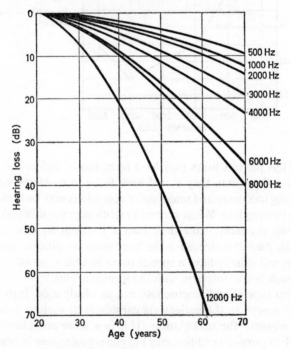

Fig. 7. Presbyacusis loss

30

women up to their mid-fifties and for frequencies up to 2000 Hz. At greater ages and higher frequencies the curves apply to women, and men on the average will suffer greater losses presumably owing—as just mentioned—to noise exposure. It is seen that the main characteristic of presbyacusis is a loss of sensitivity at the higher frequencies. Thus, for example, at the age 60 the loss at 500 Hz will, on the average, be 7 to 8 dB, while at 12 kHz it will be nearly 70 dB. It is obviously important when assessing hearing losses due to exposure to noise to allow for presbyacusis.

Tinnitus

Most people at one time or another have experienced a 'ringing' or some other kind of noise in their ears; this is tinnitus. It appears in many forms, the most common being a fairly high frequency, not quite pure tone which comes on suddenly, lasts for a few seconds and then dies away. Low frequency tones are less common but do occur. There are many causes for tinnitus which need not concern us. It is mentioned here because there are occasions when sufferers from continuing tinnitus blame the noise on some external source, such as a factory.

Deafness

The term deafness is used here for the hearing loss due to accidents, disease or exposure to loud noises, and excludes presbyacusis.

There are three main types of deafness: conductive, nerve and central. Central deafness is due to a defect in the brain centres; it will not be discussed further here. Conductive deafness and nerve deafness may occur singly but may also occur together, providing what is called a 'mixed' type of hearing loss.

Conductive deafness is due to defects in those parts of the ear—namely the external canal, the eardrum and the ossicles—which 'conduct' the sound waves in the air to the inner ear. Examples of conductive deafness are a thickening of the eardrum, a stiffening of the joints of the ossicles, or simply a blocking of the external canal, usually by wax. This type of deafness either affects all frequencies more or less equally or, more typically, is severer at the low and middle frequencies. It is limited in amount because, even if the normal channel is completely inoperative, some sound will still reach the inner ear by conduction through the bones of the head.

A common form of deafness is otosclerosis, in which a disease of the bone makes the stapes (one of the ossicles) immobile. The deaf-

31

ness increases over a period of years as the disease progresses. It can be alleviated by an operation.

A perforation of the eardrum may be caused either by disease or by accident, e.g. an explosion. This results in some loss of hearing, but often the eardrum will heal and hearing will return to normal. In severe cases it is possible to replace the damaged eardrum by an artificial one.

Nerve deafness (or sensory-neural deafness) is due to loss of sensitivity in the sensory cells in the inner ear or some defect in the auditory nerve. There is no medical remedy for it. This type of hearing loss is usually different for different frequencies and nearly always is greater at the higher frequencies, i.e. above 1000 Hz, than at the lower frequencies. It thus resembles presbyacusis. The hearing losses caused to people working for periods under noisy conditions (see below) are of this type.

Deafness Caused by Noise

If a noise is loud enough it will immediately damage the ear-drum, perhaps so badly that it will not heal, or destroy the delicate ossicles. It is not known with any accuracy how loud a noise must be to cause such serious damage because, naturally, experiments are not made on human beings, but from experiments on animals and from such accidents to humans that have been investigated, it appears to be of the order of 150 dB.

Of more general importance is the hearing loss caused by exposure over periods of time to the noise levels that often occur in modern society. Exposure to these noises causes nerve deafness, i.e. damage to the sensory cells of the inner ear, or possibly to the auditory nerve. No surgical remedy is possible. Like presbyacusis, nerve deafness due to loud noise tends to affect first the higher frequencies (i.e. above 1000 Hz) but will often, if severe enough, extend over the whole frequency range.

A temporary loss may be caused by noise but if the noise is not too loud and not too prolonged the person will, in time, completely recover his hearing. The period of recovery may be anything from a few seconds to months, depending on the nature of the noise exposure and on the person himself. However, we are concerned here only with permanent hearing loss.

The establishment of safe maximum noise levels that people may be exposed to without risk of permanent hearing loss has been difficult because of the many variables involved. Amongst these variables might be mentioned: the loudness and nature of the noise; whether exposure is intermittent, e.g. a period of exposure and then a period of rest, or continuous throughout the working day; the previous exposure of the person to noise; and the different physical reactions of different people to the same noise. Despite these difficulties standards have been proposed, and are described in Chapter 7.

2

The Behaviour of Sound in Rooms

We have seen in Chapter 1 that when a sound is originated from a point source in air a series of sound waves proceed outwards from the source in ever-increasing concentric spheres, and that the energy at any point becomes progressively weaker as the distance from the source increases. Eventually, in the absence of any obstructions, the sound becomes so small as to be negligible. Let us now consider in what way this behaviour is modified when the sound source is confined in a room. However, it must be remembered that sound is a wave motion, and a room's shape and surfaces act on the sound waves in such a complicated away that an exact understanding of what is happening in any real room is virtually impossible. So in order to be able to produce any practical advice we have to use what are loosely called geometric and statistical methods, as follows.

GEOMETRICAL STUDY OF SOUND IN ROOMS

The analysis of the paths of the sound waves can be used to find out the path of the direct sound and what kind of distribution the first few reflections of sound will have, or to control these reflections by changing the shape, attitude and absorption of the boundary surfaces. In theory it is possible to study the path of the waves during the entire time of their travel over distances of a few hundred metres. During this time they will have been reflected from the various room surfaces many times, the number depending on the size of the room. In practice the task of plotting the path of the waves beyond the first one or two reflections is too complicated to attempt, and in any case what happens to the wave paths after they have been reflected a few times is not of great importance as will be shown later.

Reflection of Sound

When the sound waves strike one of the room boundaries some of the energy is reflected from the surface, some is absorbed by it and some is transmitted through it. For present purposes the transmitted sound is negligible, and leaving aside also, for the moment, con-

34

sideration of the absorbed part of the sound, let us consider the nature of the reflection. Suppose that the surface encountered is rigid, flat and smooth, we can liken the effect to that which occurs when a light ray is reflected from a mirror. If this comparison is valid we would expect the reflected ray to form as shown in Fig. 8 as

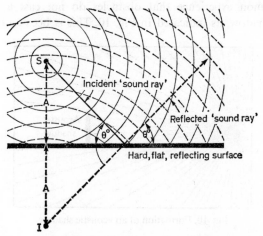

Fig. 8. Reflection from a plane surface

though it were produced by a new source (an image I) which is in a position corresponding to the original source but situated behind the surface. This illustrates the familiar law of reflection from which it is seen that the angle of reflection in all cases equals the angle of incidence.

The nature of reflections from curved surfaces can be determined

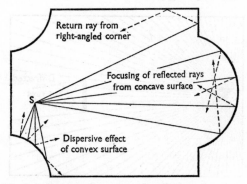

Fig. 9. Reflections from curved surfaces and corners

by the simple application of geometrical laws. The very different form of the reflected waves from concave and convex surfaces and the effect of right-angled corners between two flat surfaces are illustrated in Fig. 9.

Diffraction

It is common experience that obstacles do not cast a complete acoustic shadow, as indicated in Fig. 10. This is because of diffrac-

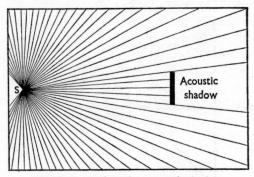

Fig. 10. Formation of an acoustic shadow

tion, which causes the sound to be bent round the corner as shown in Fig. 11. This effect occurs in light as well as sound, but its presence in the former case is not so immediately obvious, and can therefore pass unnoticed.

The reason for this is connected with the relative wavelengths of light and sound. Those for light are in the range between about 0·0004 and 0·00075 mm, compared with which the obstacles which cast shadows are very large, but the wavelengths of sound are in the

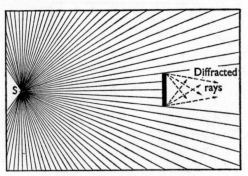

Fig. 11. The bending of sound

range of about 17 mm to 11 m, compared with which the obstacles commonly encountered range from fairly large to quite small. It should be remembered that diffraction will be more marked for the low-frequency sounds with long wavelengths than for the short-wavelength high-frequency sounds. For example, consider the effect of a sound wave encountering a 0·5 m wide pier in a room. Sounds of long wavelength, that is much greater than 0·5 m, will diffract round the pier so that there will be practically no acoustic shadow behind it. On the other hand, sounds having wavelengths much less than 0·5 m will not be diffracted and the pier will throw an acoustic shadow so that little of this sound will be heard on the far side of the pier.

These circumstances also influence the manner in which reflected waves form, the shorter waves behaving more nearly in the simple geometrical way of light waves as previously described, and the longer waves departing quite widely from this rule.

All the same, the geometrical study of sound paths in rooms is a valuable guide to practical acoustic design, for it must be remembered that the shorter-wavelength sounds play a very important part in the proper hearing of both speech and music.

Room Resonances

While the sounds with the shorter wavelengths behave like light rays constantly reflected and re-reflected until they finally die away, the longer-wavelength sounds behave in a rather different way. When the wavelength of sound is the same as one of the room dimensions, a 'standing wave' can be set up in this dimension, in which case the sound in the room behaves rather like a pendulum whose rate of swing is determined by the length of the room and whose motion tends to persist for a long time. This same effect also applies when the wavelength is exactly one-half (when two waves fit into it), one-third (when three waves fit), one-fourth, etc., etc., of the room dimension. It also applies to each of the three main room dimensions and to some extent to any subsidiary dimensions between facing flat surfaces (such as the sides of a recess) and even to the diagonals of the room.

For a room of any given size there are thus a number of sound frequencies which receive, as it were, preferential treatment, and if a complex sound containing many different notes is introduced into the room, these particular notes will be accentuated while the sound is present, and may tend to die out more slowly than other notes when the sound ceases. These resonances can be calculated from the parameters of the room sizes. If L, W and H are the length, width and height (in metres) respectively of a rectangular room, the resonant frequencies f are obtained from the formula,

$$f = \frac{c}{2}\sqrt{\left(\frac{p}{L}\right)^2 + \left(\frac{q}{W}\right)^2 + \left(\frac{r}{H}\right)^2}$$

where c is the velocity of sound (340 m/sec, approx.) and p, q and r are integers, which specify the mode of vibration. For each mode of vibration (which signifies whether no waves, one wave, two waves, etc. etc., are fitting into the particular room dimension) the integers 0, 1, 2 etc. are substituted for the values p, q and r and all the permutations completed. For example, the first few values (to the nearest Hz) for a room about 3 m cube (i.e. L, W and H are all about 3 m) are as follows:

p	q	r	Resonance f	p	q	r	Resonance f
1	0	0	55	3	0	1	170
0	1	0	55	3	1	0	170
0	0	1	55	0	1	3	170
1	1	0	78	0	3	1	170
1	0	1	78	1	0	3	170
0	1	1	78	1	3	0	170
1	1	1	94	2	2	2	183
2	0	0	110	3	0	2	198
0	2	0	110	3	2	0	198
0	0	2	110	0	2	3	198
1	2	1	132	0	3	2	198
1	1	2	132	2	0	3	198
2	1	1	132	2	3	0	198
2	0	2	154	2	1	0	206
2	2	0	154	2	0	1	206
0	2	2	154	1	2	0	206
2	1	2	165	1	0	2	206
2	2	1	165	0	1	2	206
1	2	2	165	0	2	1	206

There is thus a very large number of these resonances, many of which occur in groups at one frequency. When all the room dimensions are at least as large as the wavelength of the lowest frequency of sound, that is about 10 m, the resonances lie very close together in frequency, and there is little danger that the sound will be modified noticeably by their presence. In smaller rooms where some of the dimensions are less than about 10 m, it is found that the resonant frequencies become widely spaced out at the low-frequency end of the range. This is particularly so if two or more of the room dimensions are the same, or are related by simple ratios such as 2:1 or 3:1. The example given above clearly shows this effect. Had the three room dimensions been different from one another many more different frequencies would have appeared in this table.

When widely spaced frequencies occur sounds can occasionally be strongly influenced by them. When, for example, a single note is sounded at a frequency close to but not exactly the same as one of the strong room resonances and it is then cut off, the reverberation will change in pitch. The sound energy has in fact resolved itself into an oscillation at the frequency of the resonance of the room. If the original sound lies about equidistant in frequency between two room resonances, then the reverberation may assume a 'vibrato' effect, comprised of the two resonant frequencies with superimposed fluctuations at the rate of the difference between them.

Most auditoria are large enough so that there are many thousands, if not millions of these room resonances in the audio-frequency range, and as we have already said, there will usually be a lot of room resonances at any one frequency, and while some of them aid each other, others cancel each other out. So the overall effect is as if there were very many fewer resonances and in fact the average spacing between room resonances is related solely to the reverberation time of the room (see p. 42), and is given by $\dfrac{3\cdot91}{\text{reverb. time}}$ Hz.

STATISTICAL STUDY OF SOUND IN ROOMS

Sound Absorption

The nature of the surfaces on which the sound wave falls determines how much will be absorbed. Broadly, hard rigid non-porous surfaces provide the least absorption (or are thus the best reflectors), while soft porous surfaces and those which can vibrate absorb more of the sound. When sound energy is absorbed it is converted into heat energy, although the amount of heat is very small. As the pressure of the air momentarily increases or decreases at the surface of a porous material due to the arrival of the sound waves, air flows into or out of the pores and the friction set up between the molecules of air moving in the restricted space of the pores changes some of the sound energy into heat. Alternatively, in the vibrating type of absorbent the surface is set in motion by the alternating air pressure and the friction between the molecules of the vibrating material creates heat.

The efficiency of the absorption process is very simply rated by a number, the absorption coefficient of the material, varying between zero and one. If no sound at all is absorbed (an event which never occurs in practice) the sound absorption coefficient would be 0. If all the sound is absorbed, the coefficient is 1·0. Similarly, if three-quarters of the sound is absorbed the coefficient is 0·75. Absorption coefficients may be specified as those for sound arriving at all angles of incidence and those for sound arriving only at right angles (normal) to the surface. The first of these coefficients is the one most used in architectural acoustic design; it is sometimes referred to as the coefficient obtained by the reverberation chamber method of measurement. The coefficient for normal incidence is a more exact scientific property of the material which can in turn be derived from the acoustical impedance. These quantities enable very accurate computations to be made of the acoustical properties of a room, using wave theory, but the mathematics involved in this approach are far too complex for practical use.

Absorption of any material is not constant at all parts of the frequency scale. Indeed, the coefficient for a given material may easily be eight or nine times as great at one part of the scale as at another. Nor is the amount of effective absorption dependent only on the absorption coefficient. It depends to a slight extent on the position of the absorbent material in the room and its relation to other surfaces. For example, if a certain amount of absorbent material is fixed in patches mingled with areas having reflective characteristics, they will be slightly more effective than if the same amount of material were fixed all in one area. This is because sound waves arriving at the junction between an absorbing and a reflecting surface are bent in (diffracted) towards the absorbent material. The net effect is that the edges of absorbent material are more efficient than the middle.

Build-up and Decay of Sound in Rooms

Consider what a listener in a room receives from a short burst of sound from a source, e.g. a musical instrument, somewhere else in the room. The first sound to arrive at the listener will be the sound that has travelled directly to him from the instrument without being reflected or deflected by anything on the way, and this is known—obviously—as the direct sound. Then, in nearly all rooms, this direct sound will be followed shortly afterwards by reflections of the sound coming from various room surfaces and the amplitude of these reflections and how soon they arrive after the direct sound is determined by the shape of the room surfaces and by their absorption. In general, the time interval between these early reflections is in the order of milliseconds, because the differences in path lengths are short; e.g. the direct path length between the instrument and the listener may be say, 30 m, and the path length between the instrument

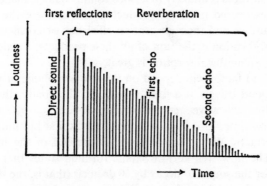

Fig. 12. Build-up and decay of sound

and the listener via the reflection off the platform floor may be, say, 31 m, so the reflection off the platform floor will arrive $\frac{1}{340}$ second later (the speed of sound being 340 m/sec.), i.e. about 3 milliseconds later. As time goes on there will be more and more reflections arriving at the listener and all—in general—getting steadily less loud ('echoes' are an exception and will be discussed later) because they have had to travel longer distances round the room to get to the listener and will have been reflected off partly absorbing surfaces on the way. This is the general reverberation process and the whole process is illustrated in Fig. 12. The time taken for the sound to die away is known as the reverberation time, and as it is one of the most important factors in the acoustical design of auditoria, the next section is devoted to a description of the process.

The Reverberation Time Calculation

Each time the sound waves meet the boundary surfaces of the room, some part of their energy is absorbed, while the remainder is represented by the waves reflected from the surface. These reflected waves also eventually meet a boundary and again a part of their energy is absorbed and a part re-reflected and so on. In the lack of continuous replacement of the original sound energy we would therefore expect any sound produced in a room to die away slowly to inaudibility, rather than to cease abruptly when the supplying energy is turned off. How long this dying-away process will take must depend on two factors, namely how much absorption occurs when the waves meet the boundaries and how often they do so. As each part of all the boundary surfaces is unlikely to absorb sound equally, the first factor must be discovered from the product of the areas of the different types of surface and their absorption efficiency or coefficient. Then the total absorption is the sum of all these products; the decay time will be less when the absorption is great.

The second factor will depend on the size or volume of the room because sound travels at a fixed speed, and the greater the volume the less often will waves meet absorbing surfaces and the more will be the decay time. This basic relationship was first put into a quantitative form by W. C. Sabine towards the end of the nineteenth century. The now well-known Sabine formula states that the time required for the sound to decay by 60 decibels (that is, the reverberation time) is found from the equation

$$RT = \frac{0 \cdot 16V}{A}$$

where RT is the reverberation time in seconds,

0·16 is a constant,

V is the volume of the room (in cubic metres),

A is the total absorption in sabins* (in square metres).

(This is illustrated in Fig. 13.)

The total absorption (A) is found by multiplying each individual area by its absorption coefficient and adding the whole together—mathematically expressed thus:

* W. C. Sabine called the units 'open window units' because they are the equivalent in absorption to a similar area of open window, from which, of course, no reflection can occur and hence has a coefficient of 1·0. They have since been renamed 'sabins' to commemorate the name.

$$A = \sum s_1 \, \alpha_1, \, s_2 \, \alpha_2 \ldots \ldots \ldots s_n \, \alpha_n$$

where $s_1 \ldots \ldots s_n$ are the areas in (m²)

$\alpha_1 \ldots \ldots \alpha_n$ are the absorption coefficients.

The reason for the specification of the reverberation time as being the time required for the sound to decay by the particular amount of 60 dB (or to one-millionth of its initial intensity) is merely to regularise the quantity for reference purposes. The only significance in the choice of this particular amount of sound decay is perhaps that it represents the range between a fairly loud sound (say a person speaking in a raised voice at about 1 m distance) and a fairly low background noise at which such a sound would become virtually inaudible.

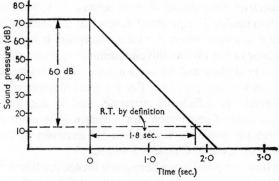

Fig. 13. Idealised decay curve

We can test the equation at its extremities to ensure that it obeys a fundamental law. Suppose we assume a room in which all the surfaces are perfectly reflecting, i.e. have a coefficient of zero. Then the reverberation time will be infinite, and since we have assumed that there will be no absorption of the sound wave in travelling through the air, this is the correct answer. In fact a very small amount of attenuation of the sound wave does take place in its passage through the air, but the amount is quite insignificant for practical purposes except at very high sound frequencies. In very large rooms (where there is a lot of air) this small extra absorption can be added thus:

$$RT = \frac{0 \cdot 16V}{A + xV}$$

where x is the air absorption coefficient, and V is the room volume in

43

m³. This is slightly dependent on the temperature and humidity as well as sound frequency, but for practical purposes these variations are too small to matter; values are given on p. 283. The value of 0·02 at 4000 Hz shows that the amount of absorption will be quite small except in rooms of large volume.

If we now test the Sabine formula at the other extremity, and assume a room having all its surfaces perfectly absorbing (co-efficient 1·0) we then obtain an anomalous result. For example, assume a room of 3 m cube, then $V = 27$, $S = 54$, $\alpha = 1·0$

$$RT = \frac{0·16 \times 27}{54 \times 1·0}$$
$$= 0·08 \text{ sec.}$$

But reverberation time should be zero, because no sound is reflected from the boundaries. For practical design this discrepancy is imma-terial except in rooms where the total absorption is very great (i.e. average α approaches 1·0) but this circumstance applies only to rooms such as some studios and to certain very 'dead' rooms used for scientific measurement purposes. For the great majority of cases the Sabine formula is sufficiently accurate, and is accepted as the standard method of calculation, for instance in examinations set for architectural qualification. Worked-out examples of the use of the formula are given in Chapters 3 and 9.

Two or three proposals have been made for the modification of the Sabine formula to overcome its anomalies. Of these, the modifica-tion due to Eyring is the one which is most often used. This is in the form:

$$RT = \frac{0·16V}{S(-\log_e (1 - \alpha)) + (xV)}$$

The term xV is the air absorption and can be omitted except for very large rooms and high frequencies as explained above. The term α is the average absorption coefficient of all the surfaces of the room and is obtained by working out the total absorption as described above and dividing by the total surface area, thus:

$$\alpha = \frac{\Sigma S_1\ \alpha_1, S_2\ \alpha_2 \ldots \ldots \ldots \ldots S_n\ \alpha_n.}{S}$$

$\log_e$ is the Napierian or hyperbolic logarithm (actually 2·3 times the common logarithm to the base 10). As we have said above, there is no need to use this formula in preference to the Sabine formula except when α exceeds about 0·25. A worked-out example is given in Chapter 9.

All reverberation formulae embody the assumption that the sound energy is evenly spread over all the surface area of the room and that the sound is perfectly diffuse. In practice this is rarely completely true but for most practical purposes we can assume that the sound field is sufficiently diffuse for the reverberation time calculations to have meaning.

The Subjective Effects of Build-up and Decay

When a steady sound is started in a room the sound level does not rise instantaneously to its final value, but builds up over a certain time. The sound level is at first due to the direct path of sound from the source. After this a further increase takes place but at a diminishing rate owing to the build-up of the reverberant sound in the room. As the direct sound level is often only a little less than the total and final sound level the subjective effect is that sound onset is almost instantaneous, the slight and slower stages of build-up often passing unnoticed. However, as we have already seen above, when the sound source is turned off an appreciable and usually noticeable time elapses (again depending on the reverberation) before the level falls to zero. This means that the hearing of transient sounds (i.e. those having sudden starting or stopping characteristics) will be somewhat modified by the presence of reverberation. Speech and music consist very largely of transient sounds and if we are to preserve intelligibility of speech and natural quality of music these transients must not be seriously affected. The transients will, of course, be modified to some extent in any practical room, but provided the modification does not exceed certain limits, all will be well. Otherwise it would be impossible to communicate by speech or listen with appreciation to music except in the open air.

Briefly stated, it is necessary that the reverberant energy from one transient sound shall have fallen to a low enough value not to obscure or mask the appreciation of the following transient. Now speech transients, generally syllables or consonant sounds, follow one another at a rate of about 10 to 15 per second and music sounds at a rate which depends on the music but can be as high as 20 notes a second. It is therefore necessary that reverberation shall not be so long nor have a character which will cause these quickly following sounds to become blurred together.

45

We have assumed up to now that a sound will decay regularly, as shown in Fig 13, but it sometimes happens during the course of the reverberant decay that an echo occurs, i.e. owing to several reflections from different surfaces arriving at the listener at the same time, or to focusing of sound, at one or more instants the level of sound rises above the general reverberant level (see Fig. 14); and the more this instantaneous level sticks up above the general reverberant level, the more likely the ear is to distinguish it as a discrete echo, i.e. a repeat of the original sound which is heard as a separate entity, and the more disturbing it will be. The double slope decay curve, illustrated in Fig. 14, is a type frequently found where two spaces (or rooms) are joined by an opening, and one of them has a much longer reverberation than the other. Alternatively, it may result from a room resonance which is insufficiently damped, the first steep part of the curve representing the average rate of decay, and the second part the rate of decay of the room resonance. In either case this type of curve can lead to undesirable results. Because the absorption coefficients of

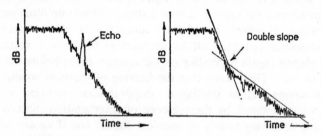

Fig. 14. Actual decay curves showing (a) an echo and
(b) double slope

different materials vary quite widely with the sound frequency, the reverberation time is liable to vary. If extreme variations are permitted it is possible that the long reverberation (and the loudness) from a sound of one frequency will obscure the perception of the following sound which may well be of a different frequency, loudness and reverberation.

For the foregoing reasons it is the general aim in acoustic design to provide a certain amount of reverberation, to ensure that this value is roughly constant over the whole audible range of frequencies and that each decay takes place at a reasonably steady rate without excessive fluctuations. Lastly, everything must be done to allow the direct sound which will not be modified by the room acoustics to reach every listener at the greatest possible strength. There is a further discussion on the subjective importance of reverberation for music in Chapter 4.

Characteristics of Sound Absorbents

Every surface of a room, of whatever material, and all the objects in it, whether animate or inanimate, will absorb sound in some degree. Many of the usual surfaces, such as plastered brick walls, windows or wood-block floors, do not absorb sound very well. In practice the actual absorption characteristics of many of these common materials depend on the way in which they are used. For example, the plastered surface of a 450 mm brick wall will have a rather different absorption characteristic from a plasterboard ceiling nailed to joists. Also there are very numerous slight or major variations in building techniques and often no data are available either because no one has troubled to measure the coefficients or more often because it is impractical to do so.

A list of approximate absorption coefficients for various common building finishes is given in Appendix A. Sound absorption is not an intrinsic property of a material alone and for this reason the values given should be regarded as representative rather than precise.

Turning now to materials having high absorption, that is, those materials which we may introduce intentionally into a room to correct or modify its reverberation characteristics, it is clear that a complete specification of absorption coefficient can only be expressed in the form of a graph (e.g. Figs 15 and 16). The first of these (Fig. 15) shows the performance of a typical porous absorbent, 25 mm mineral wool mounted directly against a solid hard surface. This shape of absorption characteristic is common to the great majority

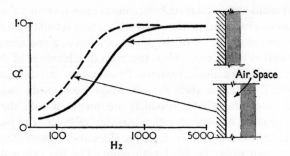

Fig. 15. Absorption characteristics of porous absorbents

47

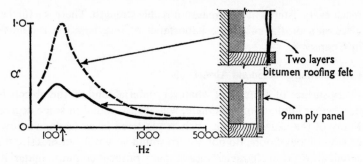

Fig. 16. Absorption characteristics of panel absorbents

of proprietary acoustic treatment materials, such as acoustic tiles, and to many of the fortuitous sound absorbers such as curtains, furniture, carpets and people. Its main feature is the considerable reduction in effectiveness at the low frequencies. There is a direct relationship between the wavelength of the sound and the thickness of the porous material for maximum absorption. Thus, thin materials can effectively absorb only those sounds having short wavelengths. Some improvement in the lower-frequency absorption of a thin material can be obtained by spacing it away from the solid backing, and the dotted line on Fig. 15 shows the changes produced by spacing the mineral wool away from its backing. A similar change in the curve is obtained when the material is made thicker.

In assessing the probable coefficients of porous absorbents the percentage volume porosity, surface porosity and degree of interconnection between the pores are important factors. The properties influence the flow resistance, i.e. the degree of difficulty with which air flows into and out of the material, and hence the absorption. Materials in which the pores are not connected, such as some of the foamed resins and cellular rubbers, cannot have very high absorption. If a porous absorbent is decorated with a film-forming paint on the surface and many or all of the pores are closed, a considerable loss of absorption will occur. This is the main disadvantage of the use of such materials as acoustic plasters. To overcome this difficulty many proprietary materials such as soft wood fibreboards and certain mineral wool and plaster products are prepared with the surface drilled, grooved or fissured with a quantity of holes. The purpose of this is to expose a porous surface (the sides and bottom of the holes) which cannot easily be filled with paint. The flat external surface may then be finished in the factory with a surface coating which is more or less non-porous, and subsequent painting, provided that it does not fill the holes, makes no difference to the absorption. A warning must be given that a number of perforated materials are not surface-coated and a marked change in performance must be expected when eventually these materials are decorated for maintenance purposes. If a porous absorbent has to be covered, for example to protect it from damage, the covering must be thin and have an open area of at least 20%. Where the protecting material is in strips wider than about 50 mm, the open area should be at least 30%, and in any case the strips should not be wider than 100 mm.

The second type of absorbent is the panel or membrane absorber having a characteristic curve as shown in Fig. 16, which shows the coefficients for a 9 mm thick wood panel mounted over a 40 mm air space and in front of a hard backing. Any plate or layer of non-porous material mounted with an air space between it and a solid backing will operate in some degree as a panel absorbent. The main feature of the characteristic is a high coefficient somewhere in the low-frequency part of the sound range and a falling absorption at frequencies above this. The position of maximum absorption depends on the resonant frequency of the panel, which can be likened to any other mechanical system having resonance, e.g. a weight supported on a spring. In the analogy the weight is represented by the mass of the panel and the spring by the compliance of the air enclosed behind it and to some extent by the elasticity of the panel material. The resonant frequency of panels of practical weights and spacings falls within a range of 40 to about 400 Hz, and is calculated from the formula

$$f_{res} = \frac{600}{\sqrt{md}}$$

where m is the mass of the panel in kg/m², and d is depth of the air space behind it in centimetres. This relationship is displayed in a convenient graphical form in Fig. 17.

Above this frequency the necessary requirement that the dimensions of the panel shall be small compared with the wavelength of the sound is unlikely to be met. Apart from this requirement the size of the panel makes little difference to its absorption characteristics.

Measurements show that the position of maximum absorption can be affected, in stiff panels, by the elasticity of the panel material and nature of the edge fixing. For example, the calculated resonant frequency of the plywood panels shown in Fig. 16 is indicated by the

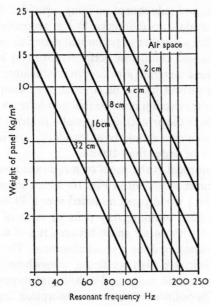

Fig. 17
Resonance frequency
of panels

small arrow. There may also be secondary peaks of absorption in the curve. However, the deviation is rarely of practical significance. When the material used is one having little or no elasticity, such as bitumen roofing-felt, then a practical curve closely approaching the ideal one is obtained. The actual measured absorption for a bitumen roofing-felt panel with the same resonance frequency is shown dotted in Fig. 16. Here the resonant peak agrees exactly with the calculated frequency. The absorption coefficient is often increased and extended over a broader range in frequency if a porous material such as glass- or mineral wool is inserted in the cavity behind the panel. This introduces increased damping by resisting the free movement of the air in the cavity when the panel vibrates.

The membrane type of absorber has application where objection is raised to the holes in many acoustic materials on the score of hygiene. Employing a fairly lightweight, supple membrane such as leathercloth over a low-density porous material such as glass-wool will enable the resonance to be raised to some frequency in the middle range, say 250 to 500 Hz, but it is doubtful if it is possible to obtain high coefficients in the higher-frequency range (1000 Hz and upwards) where much sound energy occurs in noise.

The last class of absorbent is the cavity or Helmholtz resonator.

50

This can take many forms, although all basically consist of an enclosed body of air which is connected by a narrow passage or neck with the space containing the sound waves. That an empty bottle has the properties of a cavity resonator can readily be demonstrated by blowing across the neck. This demonstration also illustrates that the resonant effect is sharply confined to one pitch of sound. The absorption provided by a single resonator (usually given in sabins per resonator) depends on a great many factors, but it can be said that in general a high efficiency of absorption is obtained at the resonant frequency but that absorption is very little for all other frequencies except those very close to the resonance point (see Fig. 18). Single-

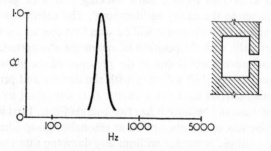

Fig. 18. Absorption characteristic of a single
Helmholtz resonator

cavity resonators can be designed to provide absorption at any point in the frequency scale, but owing to their sharp tuning they are not often used for general acoustic treatment where a major change is required, but only where a particularly long reverberation is experienced at a single frequency such as that due to a normal mode, and it is desired to reduce this without greatly affecting the average reverberation.

A much commoner application of the cavity resonator principle for sound absorption is a material consisting of an airtight plate or panel drilled or punched with a pattern of holes or slots (which form the necks of the resonators) mounted with an air space between it and a solid backing. The portion of this air space behind each hole forms the body of the resonator and in most cases there is no need to divide off the separate resonator bodies by partitions (see Fig. 19).

As in single resonators, the dimensions of the neck and body determine the resonant frequency, but the multiple resonator is not so selective in its absorption. The usual type of absorption coefficient is given in Fig. 20, which is for 10% slotted hardboard, 3 mm thick,

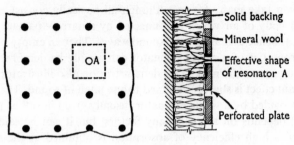

Fig. 19. Multiple Helmholtz resonator absorbent

mounted at 50 mm from a hard backing and with glass-wool or porous paper in the cavity as 'damping'. The calculated resonance point is about 1200 Hz and it will be seen that this agrees fairly well, but not exactly, with the position of maximum absorption. This shift in the resonance point is due to the presence of the glass-wool and the amount of the shift will vary with the density and properties of the porous material used in the cavity. It is important to insert this 'damping' material because it has two useful effects. First it increases the effectiveness of the absorption at resonance point (the curve for the same multiple resonator without any damping material is shown dotted in Fig. 20) and secondly it produces extra absorption at frequencies above the resonance, particularly when the holes total a substantial proportion of the whole panel area.

Taking the following practical values for the various dimensions,

Thickness of plate	1·5 to 12 mm
Diameter of holes	1·5 to 7·5 mm
Spacing of holes	12 to 100 mm
Air space behind plate	6 to 100 mm

the range of frequencies at which resonance occurs is from about 90 Hz to 4000 Hz. Thick plates with large-diameter holes, well spaced and with a large air space behind, will give the lowest frequencies. Changing any of the dimensions in the direction of the other extreme will raise the resonant frequency.

Maximum values of absorption coefficient for this type of absorber are between 0·6 and 0·9 and the coefficient is often at least one-half of the maximum over a range of at least three octaves when there is a porous material behind to provide damping. Most of the proprietary punched, drilled or slotted boards at present available have regularly spaced holes. This means that each resonator has the same frequency.

If the spacing or size of the holes were varied it would be possible to produce a material having resonators tuned to different frequencies and thereby to obtain a coefficient curve flat, or to any desired shape. Alternatively, but perhaps not quite so effectively, the variation in resonance can be achieved by varying the space behind the panel. A curve for the coefficient of the 10% slotted panel mounted so that the air space varies between 0 and 75 mm is given dashed and dotted in Fig. 20. The physical form of multiple-resonator treatments is thus capable of development into a great number of types.

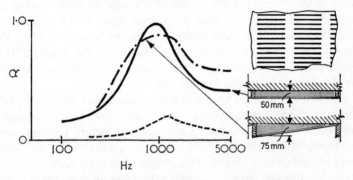

Fig. 20. Absorption characteristics of a multiple Helmholtz resonator absorbent

When absorbing materials are used to form a shaped unit, such as a sphere, double cone, cylinder or cube, the result is called a 'functional absorber'. These objects are suspended freely in a room space, some distance from the boundary surfaces, and because the sound waves in their vicinity are diffracted towards them, they provide a powerful absorbing effect, and can be used when the boundary surfaces are not available for the application of normal absorbents. Maximum absorption over a wide frequency range is obtained when their greatest dimension is between 0·45 and 0·9 m and the acoustic impedance of the surfaces is within certain limits.

3

The Design of Rooms for Speech

The main aim in designing rooms for speech is to ensure that every member of the audience can clearly hear what the speaker says; in other words, the problem is essentially one of intelligibility. A secondary aim is to preserve the natural qualities of a speaker's voice to ensure that each audience member can appreciate the nuances and dramatic effect being sought by the speaker. This requirement applies mostly to the theatre of course, and perhaps hardly at all to the public meeting hall.

The past few decades have seen a great increase in the use of loud-speaker systems for reinforcing speech, and nowadays it is usually only the smaller rooms for speech—such as lecture theatres—and the 'legitimate' theatre which do not use, as a matter of course, some form of loudspeaker system. This chapter, however, deals only with rooms for unaided speech: Chapter 5 discusses loudspeaker systems. In general it should be said that while loudspeaker systems may introduce their own problems, they can make the acoustical design of rooms for speech easier.

Nature of Speech Sounds and the Effects of Room Acoustics

Speech sounds consist of a flow of various combinations of vowel and consonant sounds. These combinations are woven into a main structure consisting of certain predominant tones which are natural attributes of the person speaking. These voice tones (sometimes called formants) can be varied over a small pitch range by the speaker, to give emphasis or shades of meaning, but are always to some extent distinctive of the person. For example, the major differ-ence between male and female speech is the pitch region in which the formants occur. The formants give the basic tone to speech and are heard most in the vowel sounds. The consonant sounds are nearly all of a transient nature; that is to say, they are very short and rapidly changing sounds, very often 'unvoiced' (i.e. containing no formant tone) and therefore having very little acoustic power. It is the correct recognition of consonant sounds which is the principal factor in speech intelligibility in an auditorium.

If every member of an audience is to hear well it is obvious that the sound at each audience seat must be loud enough. A speaker can raise his voice to adjust its loudness to suit the size of audience he is addressing but there is a limit to his capabilities, although voice training is a help. It is therefore essential that the best use is made of the limited acoustic power available, especially if the room is large, and that the background noise such as that from traffic outside the building or ventilation plant shall be kept below the limits described in Chapter 7.

Reflections of the speech sounds which arrive at the listener not later than about 35 milliseconds after the arrival of the direct sound coalesce in the listener's brain. The intelligibility of speech in a room depends basically on the speech energy arriving at the listener within this time interval being above the level of the reverberant sound; otherwise the sound of each syllable will be obscured by the still present reverberation of previous syllables. Thus the longer the reverberation time the greater the risk to intelligibility, and correspondingly the louder the direct plus first-reflected sound, the better the intelligibility will be. An obvious example is when the listener is very close to the speaker; then the direct sound is very strong and it matters hardly at all how long the reverberation time is. When the listener is some way from the speaker then the direct sound will be less due to the inverse square loss (and usually the first-reflected sounds will be less) while the reverberant sound is at more or less the same level throughout the room, so that the intelligibility will be less. However, if the reverberation time is very short, while this will not detract from the intelligiblity, the room will sound unpleasantly 'dead'.

Basic Acoustic Requirements of Rooms for Speech

We have said that it is important that the direct sound shall be as strong as possible. This direct sound weakens with distance according to the inverse square law, and therefore the average distance between the source and the listener should be kept as small as possible. It is also important that this direct sound path shall not be obstructed. This not only means that there should be no part of the building such as columns or balcony fronts interposed in the path, but also, because sound is absorbed very strongly when it passes at grazing incidence over an audience, that the seats should be arranged so that the heads of one row of the audience do not obstruct the direct sound paths to the people in the row behind. The best way of ensuring this is by seating the audience so that a clearance of at least 75 mm is provided between the sight line from one row and the sight line from the next. The clearance can, with advantage be 100 mm or more. This arrangement also provides, of course, a good view for the audience. It is axiomatic that if the audience cannot see the speaker well, there is little chance that they will hear him well. It is frequently necessary, for other reasons, to use a flat floor to the auditorium. Then the best that can be done is to raise the speaker's platform sufficiently high to ensure that the minimum clearance is obtained at the rear rows of the hall.

In plan, seats should be ranged so that none falls outside an angle of about 140° subtended at the position of the speaker. This is because speech is directional, and the power of the higher frequencies on which intelligibility largely depends falls off fairly rapidly outside this angle.

Apart from true reverberant sound, i.e. sound which has been reflected from surfaces many times, the reflections arriving at a listener's ears within about 35 milliseconds of the arrival of the direct sound operate as a contribution to direct sound. In practice the requirement is to provide hard surfaces in the room angled so that the reflected waves are directed towards the audience, and preferentially towards those members who are most distant from the speaker, since these will be in most need of some additional sound energy. Splayed surfaces at the sides and above the platform can be studied geometrically in plans and sections, and all reflected sound paths which are no greater than about 10 m more than a direct sound path from the source to the part of the audience where the reflection arrives may be expected to provide a useful addition to the 'direct' sound.

Reflected sound paths which are greater than 15 m more than the direct sound path should be avoided because such sound waves can result in echoes being heard and/or a great reduction in intelligibility. Concave curved surfaces should generally be avoided, particularly if their geometry is such as to cause reflected waves having long path differences to be focused on part of the audience or the platform. Two sections of a hall are given in Fig. 21, in the first of which are shown 'useful' reflections from a correctly designed ceiling, and in the second a bad ceiling shape which would be liable to cause a serious echo to develop. The side walls of a room may be parallel but are probably better converging on to the platform (giving a fan-shaped plan), because this reduces the average distances between the speaker and the audience. If galleries are required the clear height of the openings under their front edges should be made as great as possible, and the depth of the seating area under the galleries should not exceed twice this opening height.

If all the surfaces of the room are planned so as to provide useful reflections, that is, there is none which, because of its position or shape, is liable to cause echoes, then there is no reason why they should not all be of highly sound-reflecting materials. This would mean that the only highly sound-absorbing surface in the room would be the audience and their seats.

The reverberation time must be within certain limits, and it is proportional to room volume and inversely proportional to the amount of absorption. There is consequently a certain approximate

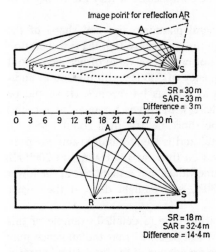

Image point for reflection AR

SR = 30 m
SAR = 33 m
Difference = 3 m

0 3 6 9 12 15 18 21 24 27 30 m

SR = 18 m
SAR = 32·4 m
Difference = 14·4 m

Fig. 21
Two long sections of halls (*a*) with usefully reflecting ceiling, (*b*) with ceiling which will cause echoes

volume of room space which, when it contains the absorption provided by one member of the audience, will have a reverberation time of about the right amount. Allowing for the fact that the room surfaces, although nominally reflecting, will have some slight absorption, this volume comes out at about 3 m³ per seat. If the above condition can be met (i.e. all surfaces useful reflectors), then it is best to design the room volume to this value, because in this way the average sound energy reaching the audience will be at a maximum. If the volume is made greater than this (and it may be considered necessary, to give good architectural proportions), then additional absorbents other than the audience will have to be introduced to achieve an acceptable reverberation time. It is therefore best to keep the volume to as near to 3 m^3 per seat as possible, and it is strongly recommended not to exceed 4·5 m^3 per seat, particularly in rooms for very large audiences.

There will obviously be a maximum size of room in which it would be reasonable to expect adequate hearing of unamplified speech. No exact value can be given for this maximum; an expert speaker in an acoustically well-designed room could certainly cope with an audience of at least 1000, but not all speakers are expert and it has become common for people to expect speech-reinforcement systems in rooms seating more than about 300 to 500 people, and even in still smaller rooms. The one exception is the theatre (see below) where the speakers should all be expert, and where speech-reinforcement systems (see Chapter 5) have to be complicated if they are to begin to be acceptable.

Having decided on the general planning, size and shape of the room, the next step is to design the surfaces so that the reverberation time will be suitable. This is not critical; a value of between about 1·0 and 1·2 seconds at a frequency of 500 Hz with maximum audience will be suitable, or even as high as 1·4 seconds. (It should be remembered we are dealing here with rooms for unaided speech.) If the reverberation time is too short, i.e. less than about 1 second, the room tends to sound 'dead' and is slightly unpleasant to speak and to listen in. The reverberation times at 125 Hz and at 2000 Hz should also be of about the same values, but a slight rise at the lower frequencies, perhaps up by 30%, and a slight drop at the higher frequency, will not matter.

In calculating the reverberation time (a detailed example of this calculation is given on p. 65) it will be noted that the audience may themselves provide a very large proportion of the total sound absorption, particularly in a large hall. This means that in the absence of an audience, or when there are many empty seats, there will be a marked change in the reverberation time. This difficulty can be partially overcome by using seats which provide almost as much absorption when they are empty as when occupied. Only well upholstered theatre-type seats will give this amount of absorption, and if other considerations forbid the use of such seats it is better to design the room for the optimum reverberation time with only, say, one-half or two-thirds of the audience present. Any type of seat, however hard, provides some absorption, of course. Values are given in the table in Appendix A.

Having decided what amount of sound-absorbent treatment is

58

required, it remains to consider where this material should be located and what form it should take. Broadly, it is best to place the greater part of the sound absorbents away from the speaker, and to apply them to those surfaces which might produce echoes, first the rear wall of the hall, particularly if this happens to be a concave curve; secondly the gallery front (if any); thirdly the ceiling margins; fourthly the top parts of the side walls and lastly the main area of the ceiling, i.e. this last should usually be kept as a reflector.

The speaker should be able to hear how the room is responding to his voice, and this will depend partly on the reverberation and partly on having reflecting surfaces near the speaker.

Absorbent materials of a hard-wearing kind, such as perforated boards or wood strips over glass- or mineral wool, should be used on walls where they are liable to suffer damage or wear. The less rugged materials, such as fibre acoustic tiles, should be used only out of reach of the audience.

We have dealt at length with the design of a large room for speech from a platform and in doing so have enumerated all the basic design problems. There are other kinds of speech rooms, in designing which all or most of the basic design requirements must be observed. There are, however, particular problems relating to some of these rooms which are described below.

Rooms for Debate

Most of the foregoing remarks have assumed that there is only one position for the speaker (this may be taken to include one area, such as a platform and not merely a single point). The requirement of good hearing from any point in the room arises in debating chambers. This makes the planning more difficult in that it is usually impossible to design reflecting 'splays' which are equally effective for sources of sound anywhere in the seating area. The volume of the room must be kept as small as possible, preferably not exceeding 2 m³ per seat. The seating must be arranged on suitable steep rakes so that each person has good uninterrupted sight and sound lines to all the other occupants. Useful reflections from the ceiling must be encouraged by suitable shape and by keeping the ceiling level not too high, i.e. not more than about 8 m.

Frequently, debating chambers are provided with public seating space. Although those seated in such areas need to hear the debate well, there is no need to ensure good hearing of sound from the public space into the actual debating area, in fact rather the reverse. The public area may take the form of one or more galleries which may be at some distance from the debating floor. It is desirable that these parts of the room, particularly if they are in the form of adjuncts, should be treated as acoustically 'dead' areas by the use of carpet (which will prevent unwanted foot-traffic noise as well as absorbing airborne sound), upholstered seats and sound-absorbent ceiling or walls or both.

Multi-Purpose Rooms

In many cases an auditorium is required to be suitable for both speech and music of one form or another. The best acoustical conditions for speech are not the same as those for music, music usually calling for a much longer reverberation time (see Chapter 4). Thus the choice of a suitable reverberation time must be a compromise. The primary purpose of the room and also tradition must influence the choice. For example, the assembly hall for a large comprehensive school will probably be used for theatrical performances, music and the showing of films, but we might concede that the primary purpose is for speech, addressed to the scholars. It is best, therefore, to aim for the ideal conditions for speech intelligibility and accept the fact that music heard in this hall will suffer from lack of 'tone'. On the other hand a church with a tradition of choral singing accompanied by an organ demands a different compromise, in fact one favouring musical acoustics.

The reverberation time of a room can be changed by making some of the room surfaces with changeable absorption characteristics, e.g. rotating panels, reflective on one side to give a longer reverberation, and absorptive on the other side, to give a shorter reverberation. However, when—as is often the case—a large part of the absorption is due to the audience, the amount of changeable surfaces which it is practical to provide may not be big enough to be worth bothering with; but each case must be considered separately on its merits. An extension of this technique is to alter the other factor which controls the reverberation, i.e. the volume of the room, by having movable surfaces, often the ceiling and sometimes also the walls.

(The problems of multi-purpose auditoria are discussed further in terms of electro-acoustic aids in Chapter 5.) It should be mentioned here that rooms such as school halls which may have reflecting, flat ceilings and floors without fixed seating are susceptible to flutter echoes when there is only a small audience (see p. 75).

Theatres

All the general recommendations given above apply to the acoustic design of theatres, and in particular it is of great importance to arrange the audience compactly so that the average distance from seating to stage is as small as possible, preferably not more than 20 m. Good sight lines are, of course, absolutely essential. The visual and acoustic requirements have without doubt influenced the development of theatre shape through the ages, and one result has been the proscenium arch type of theatre. This type can be very satisfactory acoustically if properly designed: attention is particularly drawn to the value of a reflecting splay protruding into the auditorium from the top of the proscenium opening, and indeed to the correct employment of useful reflecting surfaces in the whole ceiling design (although there may be a conflict between the acoustical and stage lighting requirements).

Emphasis is laid on the value of carpets in theatres. Not only do they provide useful sound absorption, but also reduce intrusive audience noise. Their omission is often a false economy. It is sometimes necessary to locate some absorbents on the side walls near the stage end of the auditorium. This reduces cross-reflections and reverberation in this area, both of which can adversely affect intelligibility in the front rows of seats.

A fly tower in a theatre may have a volume as great as that of the auditorium proper. The acoustical problem is that if the tower is empty of scenery, then its reverberation time may be appreciably longer than that of the auditorium, and this will be heard by the audience continuing after the reverberation in the auditorium has stopped. This may not be too serious, but probably the best course is to introduce enough absorbent into the fly tower to ensure that the reverberation time in it is a little shorter than the auditorium reverberation time. When there is scenery in the tower the reverberation time will, of course, be shorter still, and while this may make it a little dead for the actors and actresses, it will not be so bad as being much too reverberant. The calculation of the reverberation times of both fly tower and auditorium cannot be done accurately but it is sufficient for practical purposes to consider both volumes separately and to add to the area of absorption for each the area of the opening between them.

Some of the more modern theatre types, e.g. theatre-in-the-round, or thrust stages, present special acoustical problems, and these arise

61

from the inevitability of some of the actors having to have their backs (or at least their profiles) to some of the audience some of the time. The human voice is directional, more sound, particularly at the higher frequencies important for intelligibility, being emitted to the front of the speaker than to the sides or rear. It follows that when the speaker turns away, the level of the direct sound falls, as probably does also the level of first reflections, while leaving the reverberant level unchanged, and possibly leaving the level of echoes, if any, unchanged. Thus the direct sound level falls relative to the reverberant sound level impairing the intelligibility, and it also falls relative to the level of the echoes (if any) again impairing the intelligibility and making the echoes more noticeable. All that can be done to counteract this effect is to try to place surfaces so that they reflect sound to the audience deprived of the strong direct sound, but this usually runs counter to, or at least is very difficult to reconcile with, the wishes of the scenery designer.

Lawcourts

Difficulties in hearing in lawcourts are frequently reported. This is probably largely because witnesses and prisoners are naturally reluctant to speak up. The importance of ensuring that the courtroom is well insulated from extraneous noise cannot be over-emphasised, particularly when the site is surrounded by busy streets. The courtroom should be planned so as to bring the witness box and bar as close as possible to the magistrates' or judge's bench and the jury seating. Very high ceilings should be avoided, and the volume of the room and reverberation time should be adjusted to the values recommended earlier. Resilient floor coverings in public spaces and general circulation areas will ensure a minimum of foot-traffic noise.

Classrooms and Lecture Theatres

The acoustical design of small classrooms, say up to about 100 m² in floor area, depends very much on the nature of the building structure. If it is one of the recent types of building with lightweight panel walls, large window areas and a light suspended ceiling, then there will very likely be no need to incorporate any special sound-absorbent treatment, or at most a small amount of acoustic tile or some similar absorbent material at the sides of the ceiling. A rough check of the average reverberation time can be made and if this is in the region of 1·0 to 1·5 seconds (with the room empty) the conditions will be suitable. On the other hand, if the building is of heavier and more sound-

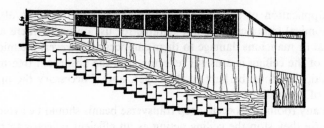

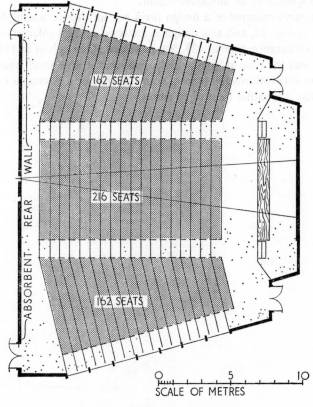

162 SEATS

216 SEATS

162 SEATS

ABSORBENT · REAR WALL

0　　　　　5　　　　　10

SCALE OF METRES

Fig. 22. A design for a lecture theatre

reflective materials, e.g. brick walls, solid concrete ceiling, etc., then in order to reduce excessive reverberation (which can rise in this type of building to a value of several seconds) some sound absorbents must be incorporated. The ceiling is the most tempting surface for the application of sound absorbents, largely because most of the economical treatments are not suitable for use on walls where accidental or malicious damage to them can occur. However, the middle part of the ceiling is a useful reflector, and the absorbent treatment should, therefore, be kept to the edges, and if necessary the upper part of the walls.

In any room for speech, deep transverse beams should be avoided, because they stop the ceiling acting as an efficient reflector to reinforce the direct sound, and because they may reflect the sound back to the speaker to an unwanted extent.

A typical example of a design for a lecture theatre seating 540 is given in Fig. 22, and this design is analysed and all calculations of the reverberation time at 125, 500 and 2000 Hz are given in Table I. Note that the time at, for example, 500 Hz is a little short when the theatre is two-thirds full (and of course is shorter still when it is full) but that it is rather long when empty.

Table 1 Reverberation time calculation for a lecture theatre
(see Fig. 22) by the Sabine formula

Surface	Area m²	Hz, coefficients and sabins					
		125		500		2000	
		α		α		α	
Floor, cork tiles on concrete including risers	400	0·02	8	0·05	20	0·1	40
Ceiling, plaster on solid	330	0·03	10	0·02	7	0·04	13
Window, 6 mm glass	30	0·1	3	0·04	1	0·02	1
Ply panels (with 25 mm mineral wool in cavities) on walls	200	0·4	80	0·15	30	0·1	20
Perforated (10%) plywood 25 mm mineral wool (rear wall)	45	0·25	11	0·75	34	0·8	36
Seats, padded	540 (No.)	0·1 (each)	54	0·15 (each)	81	0·2 (each)	108
Air	1530 m³	—	—	—	—	0·007	11
(Totals empty)			166		173		229
Two-thirds audience (extra per seat)	360	0·05 (each)	18	0·25	90	0·25	90
Totals (two-thirds full)			184		263		312
Reverberation time (seconds)	Empty	1·6		1·4		1·1	
	Two-thirds full	1·3		0·9		0·8	

4

The Design of Rooms for Music

Introductory

The acoustical environments in which music has been performed
have had a large influence on the whole art, both in the historical
development of music and in its present appreciation. Man's interest
and enjoyment in the contributory effects of reverberation have
probably derived from his centuries' long background of living in
buildings and may indeed go back to the prehistoric days when he
first took to occupying caves. This would suggest that those peoples
who have persisted in an open-air or nomadic life might be less likely
to appreciate the acoustical effects of rooms, and reflection on the
types of music made by such peoples seems to confirm this.

Certainly, the recitation of the liturgy in very large basilican
churches was influenced by the acoustical conditions; the long reverberation time made some form of intoning irresistible. Similarly, the
development of the Italian opera house with its elaborate ornamen-
tation, plush furnishings and tiers of boxes—resulting in a short
reverberation time—provided the suitable acoustical conditions for
the rapid music written by Mozart and his immediate predecessors.
(Another influence on the development of opera houses was the
tradition for attendance to be a social occasion, see Fig. 23.)

Until about the beginning of the nineteenth century music had its
appropriate auditoria—church, opera house, salon—and was written
in styles which, consciously or unconsciously, had been moulded by
the acoustics of those auditoria. However, at about that time musical
and social changes led to the present state of affairs where music is no
longer conceived in terms of the acoustical environment of any particular building but instead now demands that buildings be created
to suit its own requirements, a step first taken in practice in 1876 by
Wagner at Bayreuth. Nevertheless, the acoustical design of rooms
for music is still strongly based on tradition. This is not to say that
the tradition cannot change. In music, as in all things, a developing
process is always present. Nor is tradition necessarily of one kind,
because as we have shown music is of many kinds and has been

66

Fig. 23. Royal Opera House, Covent Garden, London

heard in a wide variety of circumstances. Added to this is the diversity of opinion found in all artistic matters, each opinion no doubt being influenced by the personal experiences of its holder.

The present state of knowledge about the acoustics of rooms for music is such that major faults (such as echoes) can be avoided in the design or, failing this, can be eliminated by suitable remedial

67

measures in the completed building. It is thus possible to design a room which has good acoustics but it is not possible to be sure of designing a room with excellent acoustics. One reason for this is that there is no firm agreement among musicians as to what constitutes excellent acoustics, and in any case acoustics which are excellent for, say, romantic music will not be excellent for classical music. Another reason is that acoustical design is a matter of compromise and the degree of compromise that is necessary will depend on the circumstances. For example, it may be necessary to sacrifice some of the acoustical qualities in order to ensure reasonable conditions for the audience at the back of the room.

Further, as will be obvious in this chapter, nearly all the advice that can be given is qualitative only, at this stage of knowledge. The one important exception is the reverberation time which can be specified (at least within a range of values) and which can be calculated beforehand with a fair degree of accuracy. Also, reverberation time can be measured accurately and easily, and until recent times is the only acoustical factor that could be so measured objectively. However, it looks as if in the next few years other objective measurements will become more common, e.g. the amount of sound energy arriving at a listener within the first x milliseconds compared with the total energy arriving. If obvious faults are found to exist in the completed building then their causes can be tracked down.

(There is no doubt that the reverberation time is a very important acoustical factor, but why it should be so important is not clear. Perhaps it is not the actual time that matters, i.e. the rate at which the musical sounds decay; perhaps it is because the reverberation time is a measure of the amount of reverberant sound (as distinct from the first-arriving sounds) reaching the listeners. If this argument is correct, then to increase the musical 'fullness of tone' (see below) in a given room, it would be no good to increase the room's volume: the amount of absorption in it would have to be reduced.)

We will now discuss the musical qualities that are desirable, and then describe the design factors that affect them. Lastly, the different classes of rooms for music will be considered.

Musical Requirements

The musical requirements (not in order of importance) that are affected by the acoustical design of rooms are: (i) definition, (ii) fullness of tone, (iii) balance, (iv) blend, (v) no obvious faults, such as echoes and (vi) a low level of intruding noise. Further it is desirable to obtain reasonably uniform acoustics over the whole audience area.

Some description of these musical terms is necessary. 'Definition' has two main characteristics: the first is concerned with hearing clearly the full timbre of each type of instrument so that they are readily distinguished one from another; the second is concerned with hearing every note distinctly so that, for example, it is possible to hear all the separate notes in a very rapid passage. (Speeds of playing of up to 15 to 20 notes per second are not uncommon.) This implies that the sounds from the whole orchestra should be heard well synchronised. 'Clarity' is a term commonly used as an alternative to 'definition'.' Fullness of tone' is the most difficult to define although it is easily recognised. Perhaps all we can usefully say about it is that it is the satisfying quality added to the sounds produced by musical instruments (or voices) when in a room as compared with in the open air. Although there may be subtle differences we must assume for design purposes that musicians mean nearly the same quality when they use such terms as warmth, richness, body, singing tone, sonority or resonance. 'Balance' we would define as the correct loudness ratio between the various sections of an orchestra as heard by the audience. 'Blend' is another quality difficult to define but in general terms it is the possibility of hearing a body of players as a homogeneous source rather than as a collection of individual sources.

All these qualities are important in all types of rooms for music but the emphasis is different. For example, fullness of tone may be more important than definition in a concert hall, while in an opera house the reverse may be true.

Design Factors

For design purposes the most convenient way of regarding the behaviour of musical sounds in an auditorium is as follows (as already outlined in Chapter 2 but largely repeated here for convenience). The sound from any given instrument or voice will spread out more or less uniformly in all directions. (Actually, most music sources are directional at some part of the frequency range. However, this need not concern us here except for one example which is discussed below.) As the distance from the source increases the loudness of the direct sound (i.e. the sound travelling directly from the source to the listener) decreases owing to the spreading of the sound (the inverse square law), and roughly speaking the loudness will be halved for every trebling of the distance. (A 10 dB reduction in a sound is roughly equal to a halving of the loudness, and trebling the distance is roughly equal to a 10 dB reduction due to inverse square loss.) Thus, for example, the loudness of the direct sound for a listener 30 m from the orchestra will be half that for a listener about 10 m from the orchestra. Further, the loudness of this direct sound will be still more reduced if it has to travel at grazing incidence over the heads of the audience: the sound waves tend to be 'sucked in' towards this absorbent surface.

All reflections of the original sound which arrive at a listener very shortly after the direct sound will to him be indistinguishable from the direct sound: the sounds will coalesce into a whole and the total loudness of the direct sound will be increased. How soon after the direct sound a reflection must be to coalesce is not known accurately, but for music it is certainly longer than the 35 milliseconds for speech (see Chapter 3), certainly up to 50 milliseconds and perhaps even up to 80 milliseconds. The speed of sound is about 340 m/s so this 80 milliseconds corresponds to a path difference of about 27 metres. For convenience we will call all the sound reflections that arrive at the listener within this 80 milliseconds the 'first-reflected' sounds, although some of them may have been reflected more than once.

After this initial period there will be many reflections of the sounds arriving at the listener and all getting steadily less loud because they have had to travel longer and longer distances round the auditorium to get to the listener and will have been reflected off partly absorbing surfaces on the way. This is the general reverberation process. Any particular reflection which is louder than its immediate neighbours in time will tend to be heard as an echo: the more it 'sticks out', so to speak, above the general reverberant sound the worse it will be.

(i) Definition

The two requirements 'definition' and 'fullness' are interrelated. Definition depends on the listeners receiving the direct sound and first reflections of it arriving not more than 80 milliseconds later at a strength well above the reverberant sound level. On the other hand, fullness appears to depend mainly on having plenty of reverberant sound. Thus it is likely that definition will suffer if the reverberation is made sufficient for maximum fullness, although there may be a range over which fullness can be varied without any appreciable loss of definition.

Another point is that the loudness of music depends partly on the direct and first-reflected sounds. Thus, if sufficient loudness is to be maintained over the whole audience area then the direct and first-reflected sounds must be maintained at as high an intensity as possible.

The audience near the front of the hall will obviously receive the direct sound loudly, but as we go towards the back of the hall the intensity of this direct sound will fall off. Nothing can be done about the loss due to the inverse square law, but the grazing incidence loss can be kept to a minimum by raking the seats. For music the 'free height' between successive rows of seats should be 100 mm if possible. Further, the longer the hall the more the back rows are going to lose definition, and it is obviously desirable to keep the length to a minimum. This is one advantage that fan-shaped or horseshoe-shaped halls have over rectangular halls.

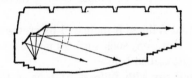

In the larger halls, say those seating about 1500 and upwards, the definition and the loudness at the back will not be very good unless the first-reflected sounds are used to reinforce the direct sound. For example, a reflector over the orchestra can be shaped so that it reflects the sound towards the back of the hall. The sounds travelling by this path will still be reduced by the inverse square law but will no longer be at grazing incidence. Thus their intensity when they arrive at the back of the hall may be the same as or even greater than the intensity of the direct sound and they will help to maintain definition and loudness at the back. Similarly, the ceiling may be so shaped as to direct sound towards the back.

If there is a balcony the listeners under it will be shielded from

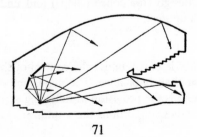

many of the useful reflections from the main ceiling, and thus any reflector over the orchestra should reflect sound into this area. Also, the soffit of the balcony should be used as a reflector.

There is a school of thought which believes that overhead reflectors make the music sound harsh. This may be so, but certainly the definition at the back of a large hall will suffer in the absence of strong first reflections.

(ii) Fullness of Tone

This quality is governed mainly by the reverberation time—the longer the reverberation time (within reason) the better the chance of obtaining adequate fullness. What the reverberation time should be is discussed below under types of hall, but fullness is also affected by the strength of the direct and first-reflected sounds. It appears that if the direct sound is assisted by strong first reflections then the reverberation time should be longer than if there were little first-reflected sound.

It is possible that fullness is increased by cross-reflection between parallel surfaces. However, such surfaces may cause resonances and flutter echoes as described below.

Probably bound up with fullness of tone is the feeling that a listener has of being enveloped in the music, (rather than it all taking place remotely from him). This feeling seems to depend on lateral reflections arriving at the listener within 80 milliseconds of the direct sound.

(iii) Balance

This quality is partly under the control of the conductor but the platform or orchestra pit design will also affect it. It is obvious that the weaker instruments such as the woodwind should not be still further weakened by, for example, being hidden behind the strings. On the other hand the strongest instruments such as the brass will come to no harm if so screened. These points are discussed further under platform design (for concert halls) and under orchestra pit design (for opera houses).

(iv) Blend

Again, this quality is partly under the control of the conductor but to a lesser extent than is balance. The main factor in room design that can help blend is the provision of reflecting surfaces close to the orchestra. Under these conditions the sounds are to a certain extent 'mixed up' before they reach the audience. Added advantages are that the players can hear their fellows clearly and that support is given to their own playing. Further, if the orchestra platform is too wide the sounds will arrive at the listeners from widely divergent directions, and this detracts from the blend. This also is discussed further below under platform and pit design.

(v) Faults

There are three acoustical faults that can ruin an otherwise good room and they are: echoes, resonances and flutter echoes.

The terms 'echo' and 'resonance' are sometimes used by musicians as synonyms for reverberation, but here by 'echo' we mean a repeat of the original sound coming so loudly and so long afterwards that it is heard as a separate entity, and by 'resonance' we mean the accentuation of a small frequency band of sounds. This latter will cause these frequencies both to be louder and to die away more slowly than the rest of the frequency range.

Thinking in terms of geometric acoustics, an echo is caused by a reflection from some surface. The seriousness of an echo is determined by how long it is after the original sound and by how loud it is compared with the original sound—the further behind it is and the louder it is the worse it will be. Two other points are, first, that the longer the reverberation time the less serious any particular echo is likely to be, because it will tend to be 'covered up' by the reverberation; secondly, that the higher-frequency echoes are likely to be more serious, mainly because the ear is more sensitive to echoes at the higher frequencies but partly because the shorter the wavelength of the sound the closer it will follow geometric laws of reflection.

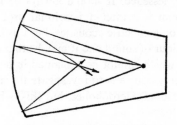

Consider for example, the reflection off the rear wall. The audience near the front of the hall will receive the original sound loudly; the reflection from the rear wall will be a good deal less intense than this because it has had to travel to the rear of the hall and back to the

C*   73 .

front again. On the other hand, for these listeners the time interval between the arrival of the original and reflected sounds will be considerable because of the long path difference (i.e. the difference between the path taken by the original sound, which is just from the orchestra to the front part of the hall, and the path taken by the reflected sound, which is from the orchestra to the rear of the hall and then back to the front). The echo will be worse, in this example, if the rear wall is concave—as it would usually be in a fan-shaped hall—because this would increase the loudness of the reflected sound at the front of the hall by focusing. By contrast, for listeners at the back of the hall the reflection off the rear wall will be comparable in loudness with the original sound but the time interval between the arrival of the two sounds will be less. If the rear wall were made absorbent then the intensity of the reflected sound would be lessened and the risk of echo from this surface reduced. (If the rear wall were perfectly absorbent at all frequencies—which is not practicable— then there would be no risk.) Similarly, if the exposed area of the rear wall, i.e. the area above the heads of the audience, is small then the intensity of the reflected sound will be less and the risk small. An alternative treatment for the rear wall would be to make it diffusing.

The rear wall has been discussed as a simple example, but obviously other surfaces of a hall may also cause echoes. Further, a twice-reflected sound (e.g. via rear wall and ceiling) might also be loud enough to cause an echo. As discussed in Chapter 2, any reflecting concave surface is dangerous but the greater the radius of curvature the less the focusing and thus the less the danger. Further, if a concave surface focuses only in space where there is no audience the danger is obviously lessened. It should always be remembered that the design of a room is a three-dimensional problem. The design methods must rely on geometric acoustics, but this is a simplification of the actual behaviour of sound and it may be that echoes will occur in completed halls which cannot be explained in terms of geometry. In such cases the only course is to investigate them by instrumental methods, e.g. directional microphones, but this is beyond the scope of this book.

The second type of fault, resonances, occur when some small frequency band is 'favoured' by the room shape. An example is two parallel and reflecting surfaces; a sound whose wavelength is exactly equal to the distance between the surfaces, or to a submultiple of it, will tend to be louder than sounds of other frequencies. (This phenomenon is known as 'colouration' in studio acoustics, and can be likened to the behaviour of a violin string which, when bowed will vibrate at a frequency determined partly by its length.)

The last type of fault—flutter echoes—is again likely to occur between parallel reflecting surfaces but the mechanism is different. In this case any short burst of sound will travel to and fro between the parallel surfaces and if the surfaces are far enough apart will be heard as a series of echoes diminishing in intensity. However, while flutter echoes will be generated between any reflecting parallel surfaces which are several metres apart if the sound is made in the area between the surfaces, when the source of sound is at some other place then the flutter echo will either be very much reduced or will disappear. For example, in a concert hall with parallel walls a strong flutter echo may be heard if the hands are clapped in the audience area between the walls but the orchestra, not being between these walls, is not likely to excite a serious flutter echo.

Both resonances and flutter echoes are only minor risks, and neither will occur if the surfaces are out of parallel by as little as 5°, or if one of them is absorbent, or diffusing. As resonances are more likely to occur at low frequencies the surface must be a good low-frequency absorbent; flutter echoes are more likely to occur at mid and high frequencies, so in this case the absorbent must be effective at these frequencies.

DESIGN OF AUDITORIA

We now give advice which is as practical as possible on the design of the various types of auditoria used for music, namely concert halls (including recital halls), opera houses, churches and cathedrals, multi-purpose halls, rehearsal rooms and music rooms. However, as we have already said and as will become more obvious, the advice must be so general and so much compromise is necessary that a good deal of judgement will be called for in any particular design. This judgement can only be based on experience and this indicates that a consultant should be employed whenever possible.

Concert Halls

(i) Shape

The three basic shapes are rectangular, fan and horseshoe. Other shapes are possible; the Royal Albert Hall is oval and a circular concert hall has been proposed but they are not common and have no tradition behind them. They may thus have faults which are not obvious on paper and which would be exposed only in practice.

Dealing first with the horseshoe shape, although this is the traditional shape for opera houses its suitability for concert halls is dubious. The main reason is that the walls must be sound-absorbent, either by using absorbent materials or by covering them with tiers of galleries or boxes, and this will result in a reverberation time too short for orchestral performances (but not for opera). Nevertheless, this shape has been successfully used in the past, e.g. the Usher Hall, Edinburgh, but then the seating was plain, thus tending to keep the reverberation time up. With the modern demand for upholstered seats it is doubtful if a long enough reverberation time can be got with this shape.

One disadvantage of the fan-shaped hall is that the rear wall, balcony front and seat risers are all curved, causing a serious risk of echoes. The rectangular hall is freer from this risk and in addition has a possible advantage that there is more cross-reflection between the parallel walls which may give added fullness. (But with parallel walls there are minor risks of resonances and flutter echoes.) The weight of tradition is on the side of rectangular halls, but then it is only in very recent times that very large audiences have had to be accommodated and the main advantage the fan shape has over the rectangular is that the length can be less. Thus the difference between the intensity of the direct sound at the front and at the back will be lessened, giving greater uniformity of acoustical conditions.

Another disadvantage of the fan-shaped hall is that it is more difficult to arrange for the lateral reflections, desirable for 'envelopment' (see p. 72).

Modifications to the basic rectangular and fan shapes are of course possible. One way to reduce a main disadvantage of the large rectangular hall—its width at the orchestra end—is to narrow in the sides at the orchestra level while keeping the full width at higher levels. Another compromise is to have a fan shape but with stepped, parallel sides.

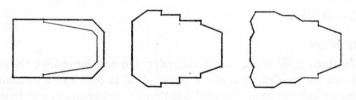

To sum up, it is probably reasonable to say that if fullness of tone is the prime requirement and that risks can be taken to achieve it, then have a rectangular shape; if definition is the prime requirement and fullness is secondary, then have a fan shape and make all the curved surfaces absorbent.

Whatever the plan of the hall the audience seating should be raked. It is better to keep accurately to the design rake than to approximate to it by one or two straight slopes. The reason is that these approximations will tend to reduce the free height for those at the back who most need strong direct sound. However, the step rise will vary and this is a nuisance and may contravene the safety regulations. (A limit will be set to the maximum rake by the safety regulations.)

The audience rake will depend on the platform height. Platform design is discussed below, but we will say here that the front part of the platform should not be less than 0·5 m nor more than 1·2 m. If it is less than 0·5 m then the 'command' which the musician—particularly the solo singer—likes to have over the audience tends to be lost; if it is higher than 1·2 m the centre section of the orchestra will be screened—both acoustically and visually—from the front rows of the audience by the front rows of the orchestra.

In a large hall some form of balcony will be almost essential—otherwise the length of the hall will be excessive. The rake of the balcony should also be properly designed. In general the audience in the balcony will get plenty of direct sound because the sound is not at grazing incidence over the stalls audience. Further the orchestra reflector (if any) and the ceiling can easily be designed to reflect sound to this area, but under the balcony the conditions are more difficult. This is because the direct sound is at grazing incidence, because this area is cut off from a large part of the ceiling and because a reflector cannot be designed to direct sound from all parts of the orchestra into this area (see p. 81). For these reasons the depth of recess under the balcony should be kept to a minimum and should never be more than twice the free height (i.e. the height between the heads of the audience and the balcony soffit) at the entrance to the recess. Also the balcony soffit should be reflecting and preferably shaped as illustrated. There is a slight risk of echo from the rear wall under the balcony, particularly in a fan-shaped hall, so for safety it should be made absorbent. However, this will make the conditions rather 'dead' for the back two or three rows of the audience and a better solution is to cant forwards the top section of the wall far enough to give useful reflections for the back rows and to prevent an echo.

This discussion of the balcony soffit profile should lead us to the consideration of the main ceiling area and the area over the orchestra, but it will be more convenient to discuss first the design of the platform. We shall see that one objective in design should be to keep the area of the platform as small as possible, so that it is better to plan for the normal maximum numbers and to leave exceptional demands (e.g. for massed choirs) to be met by temporary expedients. The normal maximum for a large hall at the present time may be taken as: orchestra 95–100, plus two pianos and four vocal soloists; choir 250–300. For all except the most important halls these figures could be reduced to 85 and 200 respectively.

The important dimensions for planning are:

1. A seated player of a violin and of most wind instruments needs an area 1 m × 0·8 m—horns and bassoons rather more.

2. A tier 1·1 m deep is sufficient for all string and wind players, including cellos and double basses; players, however, prefer at least 1·2 m.

3. Timpani and percussion need a tier 2 m deep.

4. Risers to tiers should not exceed 0·5 m because of the difficulty of carrying heavy instruments up them.

5. A piano (concert grand) measures 2·75 × 1·6 m on plan.

On the basis of these figures a reasonable allowance of space for the platform of a large hall would be: orchestra 150 m²; choir 110 m²; total 260 m².

The dangers of having a platform too large are that the sounds from the various instruments might arrive at the listeners (and at other members of the orchestra) too spread out in time, perhaps making the combined sound ragged, and not helping the blend. However, it ought to be possible to plan a platform within a rectangle of about 20 m wide by 14 m deep, and the path difference then would not be too great.

In longitudinal section an orchestra platform can be flat, flat for the front part and stepped for the back part, or completely stepped from front to back. With the flat design some of the weaker instruments—the woodwind, the violas, the cellos—are screened by the players in the front, and so are the strongest instruments—the brass and percussion—but these can stand screening. The demands of pianos have probably perpetuated what is the most frequent, but acoustically the most unsatisfactory, kind of platform—that which has a large flat area in front and a few tiers at the back. In this case only those instruments which do not need it—brass and percussion—have the advantage of exposure, while the majority of strings and wood wind screen one another's sound on the flat.

It is probable that the best design is to have a fully-stepped platform. It is true that the powerful instruments are still unnecessarily exposed, but with all the instruments given an equal chance it can then be left to the conductor to achieve the correct balance.

The platform will be made of wood on a wooden frame; this is traditional and essential for the cellos and double basses, some of whose sound is due to radiation from the platform. But it may be better to make the back tier holding the percussion of wood on solid concrete, because this will reduce their sounds a little, which otherwise tend to be overpowering. However, some works are probably intended to be overpowering, so perhaps the percussion should be given full scope; for other works the balance can be left to the conductor.

Some platform space can be saved if holes for the music stands are provided in the nibs of the risers, instead of the usual tripods. There will usually be a barrier between the orchestra and choir because when there is no choir these seats will usually be occupied by audience. (Acoustically, this is a little undesirable because they will provide absorption where none is wanted.) The barrier should be removable so as to provide greater flexibility of layout when there is a choir. The risers for the choir tiers need to be no more than 180–200 mm.

Conductors often prefer a rostrum that is off the platform altogether, and it ought to be capable of adjustment to suit individual preferences.

The placing of a pipe organ is always a problem because of its size. (Electronic organs present no problems but are often not acceptable to musicians.) The pipe organ can go behind the orchestra where it will then sound as a homogeneous whole, but with a raked platform it will tend to make the back of the ceiling too high above the main floor level and complicate the design of the ceiling area over the orchestra. If placed at one side or in two halves on either side then it will not sound as a whole, although this defect can be minimised by keeping all the pipes of a particular stop wholly to one side or the other. In either position it will provide unwanted absorption for the orchestral sounds. On balance it can be said that if some considerable sacrifice of the orchestral sound is acceptable, then the organ should go at the back; if the minimum sacrifice is required it should go at the sides. In either case the console should be movable to suit either concertos or solo recitals.

We will now consider the surfaces round the orchestra: the rear wall, the sides and the ceiling area over. In principle these surfaces can fulfil one of three purposes: to absorb the sound, to reflect it in random directions or to reflect it in specified directions. It is generally agreed that to absorb the sound is harmful; not only do the players not hear themselves but a lot of useful sound energy is being wasted which could otherwise be going to assist the direct sound in the auditorium. This is a particularly serious loss in large halls: orchestral instruments have not been developed so as to produce more sound, nor, usually are they too numerous, while audiences continue to increase in size. It seems a pity, therefore, to waste such energy as is available. Nevertheless it is often the case that the sound is absorbed, usually by curtains draped round the sides of the orchestra.

If, then, these surfaces are to be reflecting, should they be used to direct the sound in specified directions or should they provide random reflections? There is no doubt that in the larger halls—say those seating more than 2000—there is a risk that the definition and loud ness towards the back of the hall will be much less than towards the front unless some measures are taken to strengthen the direct sound reaching the back by powerful first reflections. The ceiling shape can help here (but probably cannot help those under a balcony), but a shaped reflector over the platform may be very desirable: it is a matter of opinion about balancing the loss of definition at the back against a possible harshness of tone that strong reflections from overhead may produce. Further, such directed reflections—some of which are bound to reach all of the audience—may detract from the fullness of tone unless the reverberation is made particularly long, which may not be possible. If these surfaces are made random so that the sound is diffused as much as possible then this disadvantage is lessened, but at the expense of the audience at the back. For effective diffusion, projections of the order of size of at least 0·3 to 0·6 m are necessary; an alternative is large-scale convex surfaces, in plan and section. In either case, if the surfaces are formed on frames, thus enclosing air spaces, they must be of a good weight and thickness— say a minimum of 12 mm solid wood or plaster—or else they will provide low-frequency absorption.

Diffusion round the platform also ensures that some sound is returned to the players, so that they can hear their own instruments and those of their colleagues better, and if directional reflectors are necessary parts of them should be flat for the same purpose.

It has already been mentioned that strong lateral reflections are desirable for the audience and these should arrive within about 80 milliseconds, and their direction of arrival should be more than about 20 degrees different from the direction of arrival of the direct sound.

The area over the platform either can form part of the main ceiling line or a suspended orchestral reflector can be used. The acoustical advantage of the suspended reflector is that the space between it and the actual ceiling forms a useful addition to the total volume of the hall (which must reach a certain minimum value for adequate reverberation).

Whether the ceiling shape or a separate reflector is used the requirements in a large hall are, then, to reflect some of the sound back to the orchestra to help the players hear themselves and each other—and if it has been so decided—to reflect sound towards the rear of the hall. The first requirement can be met by making parts of the area horizontal, but the second requirement is complicated by the large area covered by the sound source. It is clear that an angle for the reflecting area suitable to reflect instruments at the front of the orchestra towards the rear of the hall is not suitable for instruments at the back, and vice versa.

Consideration of the design actually used for the suspended reflector in the Royal Festival Hall will help to illustrate the problem (see Fig. 24). The height of the rear leaf of the reflector in this case was set by the demand of the organ consultant for a free opening for the organ 10 m high. This leaf reflects sound from the choir and the

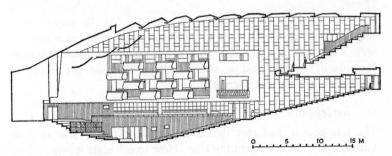

Fig. 24. Longitudinal section of the Royal Festival Hall, London

81

rear instruments of the orchestra towards the rear of the hall. On the other hand, sound from the front instruments is reflected straight back to them. If the angle of inclination to the horizontal of this leaf were increased to reflect the front sound more towards the rear of the hall, not only would the rear instruments be reflected to the ceiling but the other leaves of the reflector surface would have to be higher since the back point is fixed in space. The front leaf reflects the front instruments' sound towards the rear but the back instruments' sound towards the ceiling; a reduction in the angle of inclination in order to help the back instruments would tip the front instruments' sound too much towards the floor area.

There is a further consideration: while at middle and high frequencies the dimensions of each leaf of the reflector are large compared to the wavelengths and can therefore be considered as separate reflectors, at low frequencies the whole reflector must be considered as one. Thus if the angles of inclination of the leaves were such that the whole formed a concave surface, some focusing of the low frequencies would occur. The focusing might only occur at points in space where there is no audience, but we are not yet so confident in acoustical design as to be able to take such risks.

The reflector in the Royal Festival Hall is a compromise between the various conflicting requirements. The front leaf is at a slightly greater inclination to the horizontal than the middle leaf, which in turn is more inclined than the back leaf: the general contour is therefore slightly convex. It is made of wood 50 mm thick: the leaves are fixed by resilient mountings to timber beams which hang by tie rods from roof trusses. Its weight is 12,000 kg plus 3000 kg of lighting fittings. This reflector was made as thick as this to ensure that the low frequencies were effectively reflected. Perhaps a thinner reflector would do in many instances, but the minimum thickness should be 12 mm.

The front row of the audience should not, for the players' and singers' comfort, be too close to the platform and the space left can serve as an additional reflecting surface for the front row of players. It can be of any hard and polished material, e.g. wood, preferably on solid backing such as concrete.

(ii) Reverberation Time

The sole acoustical factor that is calculable at the present state of knowledge is the reverberation time. How to calculate it is described on p. 42 and how to measure it on p. 229; here we shall consider how long it should be and how to design a hall to get it, confining ourselves for the moment to the value at 500 Hz only.

It is clear from what has already been said that the choice of the best reverberation time is not a simple matter. First, if fullness of tone is the prime requirement then the reverberation time should be longer than if definition is considered more important. Secondly, if the hall is designed to provide strong first-reflected sounds then the reverberation time should be longer than if the platform-end surfaces are diffusive. Thirdly, romantic and choral music needs a longer reverberation time than classical or some modern music. It used to be the practice to give recommended values depending on the volume of the hall, but the modern tendency is to disregard the volume, that is, to aim for the same reverberation time whatever the size of the hall. This is a reasonable development because in the smaller hall the adverse effects of a longer reverberation, e.g. the loss of definition at the rear, are not likely to be serious. Thus a reverberation time (at 500 Hz) of between 1·7 seconds and 2·2 seconds, independent of the size, seems to be most suitable: the shorter times will give greater definition at the expense of some loss of fullness (and is the more suitable for classical music), and the longer times will give more fullness at the expense of some loss of definition, at least at the rear of the hall (and is the more suitable for romantic music).

The reverberation time is governed by the volume of the hall, by the amount of absorption in it and, to a very small extent, by the shape of the room. The audience and the seating provide an unavoidable quantity of absorbent. For example, in the Royal Festival Hall when full the audience and seating account for about 55% of the absorption at 500 Hz. In halls seating up to about 1500 there should usually be no difficulty in obtaining a reverberation time of up to 2·2 seconds. If the volume per seat is about 6 to 7 m³, then for the longer values of reverberation time it will usually be necessary to omit all extra absorbent (at this frequency of 500 Hz)—the unavoidable absorbent will be sufficient. In the larger halls, however, and assuming upholstered seats, it is probable that, while a cube per seat of 7 m³ or so will suffice to give the shorter values of reverberation time, to get the longer values the cube may have to be as much as 10 m³. This raises two problems. The first is that, to get this cube, the dimensions of the hall will be so great as to create a serious risk of echoes and of course any absorbent treatment on particular surfaces to reduce this risk will also shorten the reverberation time. One possible solution is to suspend reflecting surfaces (the so-called 'clouds') below the main ceiling over most of the audience area. These will stop one cause of echoes—the long path for sound up to the ceiling and down again—without reducing the cube per seat. But such clouds must be large compared with the wavelengths of sound if they are not to reflect preferentially, i.e. to give too much high-frequency first reflection back to the audience while allowing the low frequencies to go past them. They must be used with great caution.

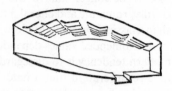

Another possibility is to make the ceiling as diffusing as possible, in both plan and section. (An example is the deep coffering in some of the excellent older halls.)

The second problem is one of cost. A hall seating 4000 at a cube per seat of 6·5 m³ will only seat 3000 at a cube of 8·5 m³. This is obvious but as the profitability or otherwise of the hall will depend on the last 20% or so of the audience there is always a powerful force at work to reduce the cube per seat.

In modern halls there is often a suspended ceiling above which there may be a considerable volume housing some of the ventilation plant, lighting fittings, etc. The absorption coefficient of this ceiling will depend on its thickness and on the volume above it, and although this coefficient may not be much the area is so large that the total absorption may be considerable. In the smaller halls, or in the larger halls when only the shorter reverberation times are desired, the ceiling absorption will not matter, but in the larger halls and when the longer reverberation times are desired, the ceiling absorption must be kept to the minimum. Little is known about the coefficients of such ceilings but it is probable that the thicker and heavier the ceiling the less the absorption: 50 mm of solid plaster is desirable.

It should be pointed out that if the reverberation time is found to be too short in the completed hall and if the absorption has already been reduced to the minimum then nothing can be done to lengthen the reverberation (except by electro-acoustic means, see Chapter 5). On the other hand, if the cube per seat is generous in the first place, not only may there be a margin in hand for acoustic treatment to stop echoes, if necessary, but if the reverberation time turns out to be too long (which is unlikely) it is a comparatively simple matter to shorten it. For example, if there is an area of wood panelling to provide low-frequency absorption some of this can be perforated to provide additional mid-frequency absorption.

If there are any considerable volumes connected through openings to the volume of the auditorium (an example would be the roof space above a suspended ceiling with ventilation and lighting holes in it) then the reverberation times of these volumes should be made rather shorter than that of the auditorium. Otherwise the reverberation will be heard still continuing in these volumes after the main reverberation has died away.

For rehearsals and perhaps for broadcasts it is desirable for the reverberation to be not too different in the empty hall than when it is full. If upholstered seats are used the difference will not be too great, but probably can be reduced further by making the underneath of tip-up seats absorbent, e.g. by perforations with absorbent behind. However, this may cause a slight increase in the absorption of the seating even when occupied and might make it more difficult to get a long enough reverberation in the full hall.

Again, if it is required to keep the absorption of the occupied seats to a minimum then only those parts of the seat which are covered by the person seated should be upholstered, leaving the sides and back hard and reflecting.

The reverberant sound should die away smoothly; this requires effective diffusion of the sound throughout the room. While there will often be enough irregularities in the surfaces to ensure sufficient diffusion, when very large plane surfaces are used opposite each other some irregularities should be deliberately introduced on these surfaces.

We have so far been considering only the reverberation time at 500 Hz. At higher frequencies the air absorption begins to become important. Further, the absorption of the audience is considerable at the higher frequencies, and the problem is usually to maintain a long enough reverberation time at this end of the scale. At the lower frequencies the reverberation time can be the same as at 500 Hz, but for fullness of tone it is almost certainly better to have it longer, say 50% longer at 125 Hz. Fig. 25 shows the recommended range of values at the lower frequencies compared with the selected value at 500 Hz.

The absorption coefficient of nearly all the surfaces in a hall and of the audience will be rather different at the lower frequencies. For

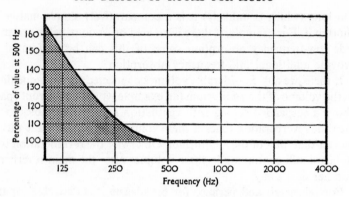

Fig. 25. Reverberation time as a function of frequency

the audience, for the seating and for all 'soft' finishes on solid backing, the absorption coefficient at 125 Hz will be less than at 500 Hz; for all 'hard' impervious finishes on air spaces the coefficient at 125 Hz will be more than at 500 Hz. While Fig. 25 shows that the reverberation time at 125 Hz may be up to 50% longer than at 500 Hz, if it is any longer than this it will be deleterious. On the other hand if the reverberation time is the same at 125 Hz as it is at 500 Hz, or is even a little less, then although we may have lost some fullness the result will not be disastrous. In traditional constructions there was sufficient low-frequency absorption present in the form of fibrous plaster or other panels with air spaces behind and of joist floors and of suspended ceilings to prevent this danger. But in modern constructions with solid walls, floors and ceilings it may be necessary to use particular low-frequency absorbents, e.g. wood panelling.

It is probable that a difference of 0·1 second in the reverberation time is the minimum perceptible when listening to music. Thus calculations should be made to the nearest 0·1 second but the inaccuracies are such that the realised value may be as much as 0·2 second different—either way—and perhaps still more at the lower frequencies.

To summarize, the normal design procedure should be to decide on the required value at 500 Hz and to choose the volume of the hall and the amount of absorption in it to get this value. Preferably, and particularly in the larger halls, the volume should be chosen so as to get a rather longer time than the design value, so that in the completed hall the time can be shortened if necessary. Then the value at 125 Hz should be calculated and if found to be more than about 25% longer than the value at 500 Hz, sufficient low-frequency absorption, e.g. wooden panels with an air space behind, should be introduced into the design. Finally, the value at 2000 Hz should be checked. Although this will be determined mainly by the air and the audience it might be possible to change some surfaces, e.g. from curtains to perforated panels, to increase the reverberation (or vice versa) at this frequency.

(iii) Faults

One of the most noticeable faults—if it occurs—is an echo. As we have said, the longer behind the direct sound an echo is, and the louder it is, the worse it will be. As a rough working rule, reflections off large flat or convex reflecting surfaces should have path differences less than about 15 m, but the real risk is from concave surfaces, which—as already said—should be avoided wherever possible.

The rear wall in any shaped hall is one of the most serious echo risks. Of course, if there is only a very small area exposed—as will often be the case in a hall with raked seating and a balcony—then the risk is not so great. Further, in these cases it might be feasible to cant the wall forward slightly so that the sounds are reflected but towards the rear few rows only.

The second dangerous area is the junction of the ceiling with the walls. The right angle thus formed can reflect sound back in the direction it came. This is particularly dangerous at the orchestra end, and in large halls one or two metres at the ceiling margin should be absorbent unless it is obscured by an orchestral reflector, or by 'clouds'.

The third area is the side walls at the orchestra end. This is one example where the directivity of instruments, in this case the brass, is important in the design. The brass emit more high-frequency sound to their fronts than they do to their sides. They will often be seated so that their sound is directed diagonally across the hall. Thus listeners near the front of the hall and seated on the same longitudinal line as the brass will receive the direct high-frequency sound rather weakly, because of the directivity, but the reflection from the side wall perhaps as strongly. The side walls can either be splayed so that the reflection goes further towards the back of the hall, thus reducing the path difference, or can be made diffusing or absorbent. Naturally. this risk does not arise in small halls where the width is such that the path difference will not be more than 10 m or so.

The last important area to consider for echoes is the ceiling. If the ceiling is shaped so as to reflect sound towards the back of the hall then the path difference is not likely to be enough to cause echoes. If, however, the ceiling, or part of it, is flat then the path difference may be too long. The worst danger is obviously near the front. Although such a flat ceiling can be made diffusing it should not usually be made absorbing because this will stop the useful reflections towards the back and will introduce too much absorbent. An alternative preventative method is to use the suspended reflecting surfaces ('clouds') which shorten the path differences without reducing the volume, but a warning of the dangers has already been given.

The other main types of fault are resonances and flutter echoes, and the only point to add to the description of these faults given on p. 73 is that parallel surfaces, which may be desirable for fullness of tone, may cause these minor faults.

(iv) Noise

The noise criteria suggested for concert halls are given in Chapter 7, and general advice on the reduction of external noise in Chapter 8. The flooring under seats should be chosen to minimise the noise of scraping feet. Carpet could be used but this will increase the absorption, probably undesirably. Cork or rubber are better because they will not increase the absorption as much.

(v) Test Concerts

When a hall is completed one objective measurement—that of the reverberation time—is possible in the empty hall, and some listening tests, e.g. for echoes, can be done. However, a full test can only be made with an orchestra and with an audience present. These test concerts are desirable because, first, some faults, e.g. a resonance between parallel surfaces, may disappear when the audience absorption is present, or may get worse, e.g. an echo, when the reverberation time is reduced by the audience: secondly, if there are any such faults they should be sought out and corrected before the hall comes into regular use; and thirdly that some subjective assessment of the acoustics can be made and perhaps some adjustments made, e.g. reduction of the reverberation time, while the builders are still on the site.

The whole audience can be given questionnaires and asked for their opinion on such matters as fullness of tone, definition, echoes, etc., but it will be sufficient if groups of about 20 people each distributed throughout the auditorium are used. In addition it may be desirable to have professional musicians, e.g. music critics, in groups and these groups should move to different positions in the hall at various stages in the concert so that they can form an opinion of the conditions throughout the hall. It will also be very useful if a few specialist listeners move about the hall, i.e. people who know something both about music and about acoustics; they may be able to decide what it is that makes one position good or another bad. For example, they may be able to decide that an echo they can hear at one position is due to a particular surface.

Opera Houses

We have said that the traditional, or Italian, opera house with its small cube per seat and large amounts of absorption resulting in a short reverberation time was eminently suitable for Mozartian operas. However, Wagnerian opera called for different acoustical conditions—more fullness of tone, less definition, greater blending of the orchestra with the singers and some subduing of the larger orchestra with its augmented brass so as not to overpower the singers. These requirements led Wagner to design his own opera

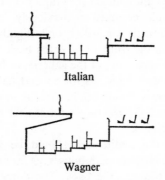

Italian

Wagner

house at Bayreuth. The main acoustical innovations were, first, the large volume per seat resulting from the absence of tiers of galleries gave a reverberation time about 50% longer than that of the average Italian type; secondly, the orchestra pit went some distance down under the stage and had only a restricted upward opening, thus blending all the instruments together and giving the correct balance between orchestra and singers (for Wagnerian operas).

There are, then two clearly defined types of opera house. Which type is chosen will obviously depend on the circumstances, or in modern circumstances a compromise may be selected. However, it should be remembered that Wagnerian and grand opera, although no doubt better with the longer reverberation time, are performed very successfully in the existing great opera houses of the world, nearly all of which are of the traditional type.

(i) Shape

The distance between the stage front and the rear seats should not be more than about 30 m (for visual as well as acoustical reasons). This means, as usual, that the design problems become worse with size: in this case when the audience is more than about 1500. The traditional shape is horseshoe and it brings the maximum numbers of audience as close to the stage as possible, but makes it difficult to get a long reverberation time. Thus it may be necessary to consider a rectangular or fan shape. The best compromise here seems to be to keep the angle of fan to the minimum with the splay continued only as far as is necessary to accommodate the number of audience. The sides near the front should be arranged to give as many early lateral reflections as possible (as for concert halls, see p. 81).

Typical proscenium widths are 10 to 18 m with 14 m typical for grand opera. Proscenium height for grand opera is about 9 m.

The seating should be raked as previously described, but as the stage platform will be flat or, better, only slightly inclined forwards, the sight lines from the front rows to the back of the stage set one limit and from the back rows of the galleries or boxes to the front of the stage set the other limit. As for concert halls the depth of the recess under a balcony should not be more than twice the free height at the entrance, and the balcony soffit should be shaped to reflect sound towards the back.

The ceiling of the auditorium should be flat or (preferably) diffusing or, in the largest opera houses, as far as possible shaped to reflect the voices towards the back. The domed ceiling has often been used in the past but produces undesirable focusing and, if high enough, echoes. This focusing can cause the loudness of the voices to vary considerably as the singers move about the stage. (There is often a conflict in ceiling design between the acoustical and the stage lighting requirements.)

It should be remembered that it is the singers, not the orchestra, who need most help and any reflecting surfaces should be designed with this in mind. A proscenium splay might be helpful, and certainly an apron stage, even if no more than 2 m from front to rear, is a useful reflecting surface. It should, of course, be boarded and not carpeted.

(ii) Pit Design (mainly after M. Barron)

An important problem in opera houses is to obtain a correct balance between singers and orchestra, and this depends to a considerable extent on the design of the pit. The traditional orchestra pit was only shallow and most of the orchestra was exposed but this may be one part of design where it may be better to depart from tradition. This is because orchestras have got larger and louder, not only for Wagner but also, for example, for Verdi. Thus it may be better to enclose— if not most of the orchestra as at Bayreuth—at least the louder sections of it. Thus the stage should overhang the pit; the top of the (solid) orchestral rail is generally in line with the stage. Fig. 26 shows the usual arrangement, typical dimensions being V 1 to 2 m, D 1 m and H 2·5 to 3·5 m.

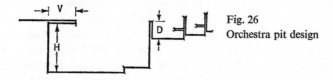

Fig. 26
Orchestra pit design

The floor area of the pit should allow for 1·1 to 1·2 m² per player, and with space for up to 100 players for grand opera, up to 120 for Wagner. The size of the pit should be made adjustable to suit orchestras of different sizes. It is becoming common practice for the depth of the pit to be made adjustable, i.e. by altering the floor height. The width should not be more than four times the depth (along the longitudinal axis). A cranked orchestra rail is often used to accommodate the orchestra in the splay in plan beyond the proscenium opening width. The interior surfaces of the pit should be reflecting but preferably of wooden panels with an air space behind; if all the surfaces are solid there is a danger of too much low-frequency reverberation.

(iii) Reverberation Time

The reverberation time for the Italian type opera house should be about 1·2 seconds (plus or minus 0·2 second) at 500 Hz when full. For the Wagner type the reverberation time should be about 1·7 seconds. It is probable that, unlike concert halls, the reverberation time at 125 Hz should be no longer than it is at 500 Hz. This is because the low-frequency reverberation will help the orchestra more than it helps the singers.

The volume of the stage (including the fly tower) is often as big as that of the auditorium. The amount of scenery and curtains in it will usually ensure that the reverberation time in the stage volume is no longer than that of the auditorium. It is desirable that it should not be any longer, and if there is any doubt additional absorbent should be introduced into it. On the other hand, it should not be made too 'dead' because the singers will need some help from local reverberation. (How to calculate the reverberation time in the fly tower is described on p. 61.) Unfortunately, because of the scenery requirements, it is not usually possible to provide what is acoustically desirable—namely, nearby reflecting surfaces for the singers.

(iv) Faults

The prevention of acoustical faults will be much the same for opera houses as for concert halls, except that flutter echoes and resonances are unlikely in horseshoe or fan-shaped auditoria. All curved rear wall surfaces which are not hidden behind the audience should be absorbent or diffusing, unless they focus in space where there is no audience. Echoes off the ceiling should be guarded against. The greatest danger is near the front and an overhead proscenium splay can be used or the ceiling can be made diffusing.

(v) Noise

The same remarks apply as for concert halls.

(vi) Test Performances

One main advantage of test performances before an opera house is opened would be to discover any faults. There is not so much need in opera houses to attempt to balance definition against fullness as there is in a concert hall but there is a need to test balance between singers and orchestra.

Churches and Cathedrals

Churches and cathedrals are nearly always used for both speech and music, and so some acoustical compromise will be necessary, and thus the acoustical design will depend on whether speech or music is considered the more important element and this will depend on the nature of the religious service. If speech is of first importance then the church should be designed in accordance with the principles given in Chapter 3. If the music is of first importance then the design should be as follows. (It should be mentioned that speech-reinforcement systems make it possible to make speech both intelligible and natural even under very reverberant conditions (see Chapter 5), although the necessary microphone technique imposes some limitations on the preachers.)

The long reverberation time (up to 7 or 8 seconds when full) of the mediaeval church led to the development of polyphonic choral music, and if this music is still to be performed then the reverberation time should be 4 seconds (at 500 Hz) when full. While this time will not be excessively long for later religious music (Bach and onwards) it would be better to have about 2·5 to 3 seconds. Also cathedrals and large churches may be used for festival music of all kinds and the shorter reverberation time is more appropriate for this. The reverberation time at 125 Hz may be up to 50% longer than that at 500 Hz.

The volume per seat will usually be large enough to give this reverberation time or longer. Any additional absorption necessary should be distributed over the walls and ceiling areas and not concentrated in one place.

A church built with separate 'cells', e.g. chapels opening off the main body of the church, has the acoustical disadvantage that the reverberation in the cells may be different from that in the main body. Thus particular notes may undesirably continue in these cells after the main reverberation has ceased.

The choir will benefit from reflecting surfaces close to them so that they can hear themselves, and the higher they are placed, the further will their sounds carry. Also this will bring them closer to the ceiling which may return helpful sound to them. The common arrangement of the two halves of the choir facing each other with solid wooden screens behind them is also helpful.

Opinions vary on the best position for the organ, but certainly it should not be far from the choir. Nor should it be narrowly confined in the organ loft. The console should be with the choir in the body of the church so that the organist can judge the total musical result. But there must be no obstruction between the organist and the organ, and from the console position he must be able to hear the full power and balance of the organ. As far as possible both choir and organ should be positioned so that their sounds have equal chances of reaching the congregation; otherwise it will be difficult to achieve good balance between them.

Extensive concave surfaces are best avoided as they cause a serious risk of echoes. If they are used then their radii of curvature should either be so large that they focus, so to speak, outside the building, or so small that they focus in space where there is no congregation. In general, the echo problem is not so critical in a church as in a concert hall because the reverberation will be longer. Nevertheless, the west wall should preferably be absorbent. We have said that any extra absorbent required to control the reverberation should be distributed over all the surfaces, but this should not preclude its use for reducing echo risks from particular surfaces.

Finally, we will mention one acoustical peculiarity and it is that a church with a long reverberation often has a 'sympathetic note'— that is to say, a region of pitch in which tone is apparently reinforced. This region lies between G and A sharp. The choir find it easier to sing in the key of the sympathetic note. Further, in many large churches where there is no sympathetic note, it is still often found that tenor A is the best note to recite on or to intone the service on. Nothing is known about the reason for this, or how to design a building to get it.

Multi-Purpose Halls

These have been discussed in the previous chapter, but they provide one of the most awkward acoustical problems because, for example, the very desirable raked floor may be impossible because the hall may be used for dancing, and there is the conflict between the shorter reverberation time desirable for unaided speech compared with the longer time desirable for music. It is usually more important to have the correct reverberation for speech rather than for music in such rooms, unless a speech-reinforcement system specifically designed for rather reverberant conditions is relied on.

The orchestra should not be confined within curtains behind the proscenium arch but instead should be brought as far forward as possible. Thus an apron stage is a big advantage. Reflecting surfaces to help the orchestra hear themselves are desirable even if they are only temporary structures of plywood. One of the worst weaknesses will be the poor hearing quality at the rear if the hall has a flat floor and stage platform. It is therefore most advantageous if an overhead reflector is used or the ceiling is shaped to reflect sound towards the rear. Such a reflector could take the form of a proscenium splay. Electro-acoustic aids for multi-purpose auditoria are discussed in Chapter 5.

Rehearsal Rooms

By this we mean rooms for orchestras or choirs to rehearse in and associated with the particular concert hall or other building. Such rooms should simulate the conditions of the auditorium proper as far as possible. Thus for a concert hall, the platform in the rehearsal room should be the same as the one in the auditorium and so should the nearby reflecting surfaces. It will not be possible to copy exactly the acoustical conditions but the reverberation time should be about the same as the full auditorium, and this will usually involve making the wall the orchestra faces at least partly absorbent. This has the added advantage that the orchestra will not get the reflections from this wall that they might like to have but which would not occur in the auditorium. Certainly the reverberation time of the rehearsal room—if associated with a particular concert hall—should not be longer than that of the hall, otherwise the performance conditions will be more difficult than the rehearsal conditions.

Obviously the volume of the rehearsal room can not be as large as that of a concert hall, but it should be made as large as possible.

Music Rooms

By this we mean rooms used for music teaching and practice. The acoustical requirements do not appear to be very exacting and a reverberation time of between 0·5 second for the smaller rooms and 1 second for the larger rooms will probably be suitable. It may be desirable to make the reverberation slightly adjustable to suit individual preferences, e.g. by having curtains which can be drawn across one or more walls.

If the rooms are of solid construction (e.g. brick walls with concrete floor and plaster or solid concrete ceilings) then the reverberation time will be too long, particularly at low frequencies. This will call for the use of absorbents, and for the low-frequency absorption wood panels on an air space would be suitable.

It is desirable that some diffusion should be deliberately introduced, e.g. by modelling of the wall surfaces.

5

Electro-Acoustic Aids in Auditoria

The last few decades have seen an enormous growth in the use of loudspeaker systems, and it is now only the 'straight' theatre, the concert hall and the opera house which do not use some form of loudspeaker system as a matter of course to help the hearing of speech or music (although the modern theatre will have a sophisticated sound-*effects* system). There are several reasons for this growth. A good system can, in many cases, mitigate or completely overcome difficult acoustics for speech, such as may occur in cathedrals; much larger rooms can be built than the unaided voice could ever cope with; owing to the radio, the cinema and television, listeners are now used to a loudness (of speech particularly) which makes listening effortless and which, in the larger rooms, the unaided voice can not achieve. Perhaps the art of public speaking—in the technical sense of making oneself heard to a large number of people perhaps under poorish acoustical conditions—is in decline, and speakers now expect reinforcement systems (although the speech training actors and actresses receive still enables them to speak, apparently effortlessly, about 10 dB louder than untrained people). And a whole class of singers have come into being, whose whole style depends on the microphone. Whatever the reasons for the growth in the use of loudspeaker systems, there is no doubt that they often have great advantages and we can expect their use to increase rather than decrease; this is why here we deal with loudspeaker systems in some detail.

Loudspeaker systems tend to have a bad reputation, and this is usually due to a neglect of the acoustical factors when installing the equipment, or to bad use, or to poor maintenance. Plenty of equipment of high quality is available, but it is of course essential to buy good quality equipment in the first place and then to have it regularly maintained.

This chapter deals more with speech systems than with music systems. This is because, in some ways, the design of speech systems is more critical, and because they are usually a permanent installation and are used by all comers. Music systems may also often be a permanent installation, but in addition several artistes carry round with them their own equipment which meets their own special needs. And even permanent installations for music will often be under the control of an expert, who will have his own decided views on what equipment to have, and how to use it, often with successful results; speech systems are more often left, so to speak, to their own devices, so it is important to get the original design as foolproof as possible, although of course speech systems should receive as much care as music systems.

In this chapter the word 'speaker' means a person speaking as opposed to a loudspeaker; a reinforcement system is defined as one that has the speaker (or singer, or orchestra) and the loudspeaker(s) (and audience) in the same room, whilst a public-address system has them in separate rooms. The acoustical design problems are the same for both, except that public-address systems are rather easier in that there is no feedback problem, and there is no need for the sound to be coming from any particular direction, which is a requirement with some reinforcement systems. So in what follows we do not distinguish any further between reinforcement and public-address systems, and the reader who is concerned only with public-address systems should ignore the points about feedback and apparent direction of source.

SPEECH SYSTEMS

Speech loudspeaker systems have to be used for either or both of two main reasons. The first is that the unaided sounds would not be loud enough: because there is too much other noise, because some of the audience is too far away or because the speaker does not speak loudly enough. These factors are interrelated, because if a listener is close enough to the speaker, the intruding noise (within reason) is not important. Put the other way round, the level of the intruding noise will determine how far away from the speaker a listener can be before the speech he is trying to listen to is lost in the noise. There is, of course, a limit to this interrelationship, because even if there were no noise at all, the speech sounds would eventually, as the listener moved further and further away from the speaker, get so quiet as to sound unnatural.

The second reason for the use of loudspeakers is that the acoustics of the building interfere with unaided speech sounds, and a loudspeaker system is required to mitigate bad acoustics.

D   97

We would define a good speech system as one with which all the audience hear the speech clearly, undistorted and at reasonable loudness. Further, in most cases the speech should appear to be coming from the speaker and not from any loudspeaker, although there are exceptions—such as political meetings—where illusion of reality is not important. Ideally, the audience should not be aware there is a loudspeaker system in use. But the most important requirement for any speech system is that it should be intelligible.

Microphones

Although we do not intend to discuss apparatus in detail, it should be stressed that good quality is essential, and this applies particularly to the microphones. If a bad microphone delivers distorted speech signals to the rest of the system, nothing can be done subsequently to put this right.

Microphones are made either equally sensitive to sounds arriving at them from any direction (omnidirectional), or more sensitive to sounds arriving from one or more particular directions. Directional microphones are nearly always used for reinforcement systems because they reduce the feedback (i.e. the howling noise produced when the amplification round the circle microphone to amplifier to loudspeaker and to microphone again is greater than unity). Feedback occurs at a particular frequency rather than at all frequencies at once because most loudspeakers vary in efficiency from frequency to frequency and because of the room acoustics. Further, poor-quality microphones will also have an irregular response. Thus feedback will happen at some frequency at which the loudspeaker and/or the microphone is most sensitive and at which the loudspeaker most easily excites a room resonance containing the microphone and loudspeaker. A directional microphone can be placed so that its least sensitive side faces the loudspeaker and thus the amplification of the system can be greater before feedback occurs, but it will be seen later how this requirement can conflict with the requirements of a high-level system (see p. 101).

Assuming that the system has been designed to provide adequate acoustic power output (which is no problem) then it is the feedback which limits the maximum loudness of the system (except of course with public address systems). If feedback is a problem, there are a few ways in which it can be reduced (some of them patented). One is to introduce filters into the circuit which reduce the gain at some of the most likely feedback frequencies (although as these feedback frequencies are determined largely by the positions of the micro-phones and loudspeakers in the room, they are liable to change if the microphones are moved). Another method is to introduce a circuit which alters the frequencies of the speech or music slightly; this evens out to some extent the feedback frequencies, but although this slight frequency shift cannot be heard with speech, it is some-times detectable—and disturbing—with music. Either of these methods enables the gain to be increased by about 5 to 6 dB before feedback occurs.

Directional microphones are used to reduce feedback. A ribbon microphone has a figure-of-eight characteristic, i.e. its sensitivity is at its maximum at the front and at the rear, and decreases steadily to zero at the sides. This shape of directivity is particularly useful when people are on either side of a table and the loudspeakers are at either end. But the most commonly used type is the cardioid microphone, with maximum sensitivity at the front and minimum (but not zero) at the sides and the back, and this has now been developed further with the hypercardioid which has a narrower front lobe than the earlier cardioids.

If the speaker is at a fixed position, e.g. at a rostrum, then micro-phone placement is not usually a serious problem. The microphone should be about 0·5 to 0·7 m from the mouth; if closer the speech gets frequency distorted, i.e. the balance of the natural voice is upset and with ribbon microphones the lower and middle frequencies are accentuated; if further away the microphone will pick up the rever-berant sound. When listening to natural speech, the brain with the help of the directional properties of the two ears can discriminate to a great extent against the reverberation of the speech. Thus the speech will still sound good up to some distance from the speaker, say, for example, up to 15 m in a hall with reasonable acoustics. The microphone has no discriminating properties, and has only to be 1·5 to 2 m away from the speaker for the speech picked up by it to sound very reverberant. The speaker must not, of course, move out of the sensitive area of a directional microphone. This is more important when a ribbon microphone is used because its sensitivity falls off to the side more rapidly than the cardioid. Nor must the speaker move about too much, and this is because the loudness will vary a lot. For example, the speaker's mouth may normally be, say 0·6 m from the microphone; if he leans forwards so that his mouth is only 0·3 m away the sound level at the microphone will go up 6 dB, and the sound from the loudspeakers will go up correspondingly. This will

sound unnatural to the audience, because this slight forward movement under natural, i.e. unreinforced, speech conditions would
make very little difference (less than 0·5 dB) in the loudness to a
listener 6 m away, and even less difference to listeners further
away.

The restriction of movement is sometimes a handicap to, for
example, dramatic preachers, whose actions are thus limited, and is
a more serious disadvantage in lecture theatres where the lecturer
wishes to continue talking while he moves over to the blackboard or
the screen. A halter microphone, i.e. a microphone hung round the
speaker's neck so that the microphone hangs on his chest, or (less
often) a small microphone fixed in the lapel, can be used to give
freedom of movement, but the lecturer must avoid getting tangled
up in the lead, and the preacher must remember to remove the microphone when leaving the pulpit.

Often, speech from several fixed positions has to be amplified.
Each position will have to have its own microphone, and it is most
desirable that only the one in use at any given moment should be
'live'. If the other microphones are left on, they pick up the reverberant sounds; this tends to confuse the speech.

It is not usual to use loudspeaker systems in 'straight' theatres,
but when they are used, the placing of the microphones is a problem.
One possibility is to use radio microphones, i.e. small microphones
worn on the actor's person and which incorporate a small radio
transmitter, the signal from which is then fed to the loudspeaker
system. This obviously calls for a high degree of maintenance.
Another possibility is to use several microphones hidden from view
and placed so as to cover the whole stage area. Because of the large
and varying distance between actor and microphone the amplified
sound tends to be reverberant and to vary in loudness. The rever-
berance can be reduced by using specially directional microphones,
and the varying loudness can be controlled by a skilled operator.
With all their difficulties, speech reinforcement systems could be a
help to overcome the troubles, already mentioned on p. 62, that
occur in theatres where the actor has to have his back to part of the
audience part of the time.

Loudspeaker Placing

It is the placing of the loudspeakers that is most often responsible for
the poor quality of speech systems. The behaviour of sound in rooms
is described in Chapter 2, and similar considerations apply when

dealing with loudspeakers. Briefly, for good intelligibility the direct sound, i.e. the sound travelling directly from the speaker to the listener, should be of adequate intensity compared with the reverberant sound. The intensity of this direct sound gets less as the distance between the speaker and the listener increases—by at least 6 dB every time the distance is doubled, and by more if the direct sound is passing at grazing incidence over the heads of the audience. On the other hand, the intensity of the reverberant sound is much the same over the whole audience area. Thus a listener close to the speaker hears the direct sound well above the reverberant sound and the intelligibility is good; a listener a long way away gets the direct sound at a much lower level and the intelligibility is worse. How much worse will depend on the reverberation time of the room and on the nature of the first few reflections of the sound, as discussed in Chapters 2 and 3. Further, the loudness must be kept at a reasonable level over the whole audience area, both for naturalness and to make sure that the speech sounds are well above any noise.

As an example we will consider a hall with a flat floor seating five hundred audience and where the reverberation time is about the optimum for speech. The distance between the speaker and the first row of listeners we will take as being 3 m, and between the speaker and the back row as 25 m. Obviously the intensity of the direct sound reaching the front row will be adequate, but the intensity of the direct sound reaching the back row will be 18 dB less (ignoring any useful first reflections), due to the inverse square law, and the higher frequencies—important for intelligibility—will be even more down possibly by another 10 dB, because of the grazing incidence over the heads of the audience. If a loudspeaker is now placed 6 m above the head of the speaker, the sound from it reaching the back row will be 12 dB less than the sound reaching the front row (26 m distant compared with 6·7 m), and the loss due to grazing incidence will be negligible. Thus the difference between front and back for the higher frequencies has been reduced from 28 dB to 12 dB (ignoring the contribution from the real voice). This is an extreme and simple example, but does illustrate the advantages of a single loudspeaker. Incidentally, this would be described as a 'high-level' system, which is the term for a system employing a very few loudspeakers each operated at a fairly high level of sound, ('level' does not refer to the height of the loudspeakers above the ground) and each covering a large number of people. A low-level system employs many loudspeakers each operating at a low level of sound and each covering only a few people. The comparative advantages and disadvantages of each type of system will be apparent later.

A listener in a hall with a speech-reinforcement system will receive the speech sounds at least twice, in the sense that he will receive the 'natural' speech directly from the speaker and also the same speech coming from the loudspeakers. The loudspeaker sound will arrive at a listener before or after the 'natural' speech depending on the relative position of the speaker and the loudspeaker, and it might be louder or softer than the natural speech, but usually louder. If there are several loudspeakers then of course a listener will receive the speech several times. The relationship in time and loudness of these repetitions of the speech sounds is most important. The permissible limits are given below, but briefly if the repetitions arrive at a listener close enough together, i.e. spaced over not more than about 35 milliseconds, then they all add together and increase the loudness and increase the intelligibility. If on the other hand they are spaced out over a larger time interval they begin to interfere with the intelligibility; the greater the time interval between them the more they interfere, and if the time interval is long enough—about 100 milliseconds—they will be heard as discrete echoes.

There is a second effect to be considered, often called the precedence or Haas effect. (Haas was the first to put this effect, and the effect of long-delayed echoes, on a quantitative basis.) If we consider now only the original sound and one repetition of it at the same loudness and following soon after (i.e. about 25 milliseconds later), it is the sound which arrives first at the listener that determines the apparent direction of the source. An example will make this clearer. A listener is seated facing a speaker 17 m in front of him, and has just to one side of him a loudspeaker. The direct sound will take 50 milliseconds to travel from the speaker to the listener, but the amplified sound via the loudspeaker will get to him practically instantaneously because it travels the 17 m along the cable and not through the air. If both sounds arrive at the listener at equal loudness, he will be conscious of only the first-arriving, i.e. loudspeaker sound. If we now introduce some form of electrical delay between the microphone and the loudspeaker, we can arrange for the natural sound from the speaker to arrive first at the listener. In our example, let the artificial delay be 70 milliseconds, then the direct sound will arrive first by 70–50 milliseconds, i.e. by 20 milliseconds. This has the very important effect of making the listener unaware of the loudspeaker, and all the sound—although in fact part of it is coming from the loudspeaker—

appears to be coming from the speaker. Thus the system is made not only more intelligible but also much more natural. This effect holds, not only when the natural and loudspeaker sounds arrive at the listener with equal loudness, but also even when the delayed sound is a good deal louder than the natural sound. In fact, the total loudness at the listener can be doubled, compared with the unaided voice, by a delayed loudspeaker without the listener being aware of the amplification (except, of course, that it is louder).

This technique can be extended if necessary over considerable distances. The unaided voice, let us say, will carry the first 15 m of a large hall and then needs help from the first loudspeaker. This loudspeaker will carry the voice at adequate loudness and intelligibility for another 15 m, let us say, towards the rear of the auditorium. Here, at 30 m from the speaker, the speech requires further amplification and a second loudspeaker can now be introduced delayed so that, to a listener close to it, the first sound to arrive is that from the speaker, the second sound to arrive is that from the first loudspeaker and the last sound to arrive, and the loudest, is that from the second loudspeaker. Acoustically, all the speech will appear to be coming from, if not the speaker himself, at least from the first loudspeaker, and as this will be roughly in the line between the listener and the speaker, it will in fact appear to be coming from the speaker.

Reliable delay units are now available which will provide any number of delayed signals at any required delay times.

A simple case involving the relative times of arrival is a reinforcement system using one loudspeaker near to the speaker. If this loudspeaker can be mounted so that it is a metre or so further from the audience than the speaker is, e.g. above his head, then the direct speech will arrive first and the amplified sound will not be noticeable. The trouble is that the dangers of feedback are increased, because the loudspeaker will be nearer to the sensitive side of the microphone than if it were mounted more towards the audience and thus nearer the 'dead' side of the microphone. Nevertheless, if feedback can be avoided it is much better to have the loudspeaker further away from the audience than the speaker.

Of course, if the loudspeaker is very close to the speaker, e.g. mounted in the front of a rostrum, then although the loudspeaker sound will arrive marginally earlier, the illusion will still be maintained.

Loudspeaker placing is discussed again later in this chapter (pp. 109 and 120).

Loudspeakers

We should now consider how loudspeakers radiate their sound. There are two main types: horn and moving-coil. Horn loudspeakers are more efficient, by about 10 dB, than moving-coil loudspeakers, i.e. they produce that much more acoustical energy for a given electrical input but, at least for speech, this is not an important advantage in auditoria because the power requirements are low, and powerful enough amplifiers are readily available; but in some industrial systems this high efficiency is useful. Sometimes they are found in high-power entertainment systems where they are often used in conjunction with moving-coil units to form a wide-band loudspeaker system. However, the moving-coil loudspeaker is the most common, and we will confine our remarks to this type.

When mounted in some conventional cabinet we can say—without going into too much detail—that at the lowest frequencies (up to about 200 Hz) these loudspeakers radiate equally in all directions; at the mid-frequencies (about 200 to 1000 Hz) they radiate rather less to the back but near enough equally over the front 180 degrees; at the highest frequencies they radiate a beam of sound (which gets narrower as the frequency increases) directly towards the front, practically nothing towards the back, and little to the sides off the beam. Thus if one of these loudspeakers is used facing towards the audience, those on the axis of the loudspeakers will get all the frequencies, but those to one side will not get the higher frequencies (which are important for intelligibility). A method of overcoming the directional effect at the high frequencies is to use one large loudspeaker to handle the frequencies up to 500 or 1000 Hz and another, much smaller loudspeaker to handle the higher frequencies. The smaller the loudspeaker the less this directional effect is, so this two-speaker arrangement with its 'cross-over' connection will not become seriously directional until a much higher frequency than would the large loudspeaker on its own. It should be explained that a small (i.e. less than 150 mm diameter) loudspeaker cannot be used on its own for two reasons. The first is that it will not radiate the lowest frequencies, and the second is that it will not handle the power that will be necessary even in the smallest hall. But used in conjunction with a larger loudspeaker, it will be adequate because only some 10% of the power is contained in the frequencies above 1000 Hz.

Properly used, however, a directional loudspeaker arrangement can be a great advantage, and the most common type is the 'line' or 'column' loudspeaker. If several ordinary moving-coil loudspeakers are mounted one above the other and all are connected together in phase, by an interference effect they concentrate most of the sound into a beam which is narrow in the vertical plane but which is no more directional in the horizontal plane than an ordinary loudspeaker. This is illustrated diagrammatically in Fig. 27. An analogy is the light from a car's flat-beam fog lamp.

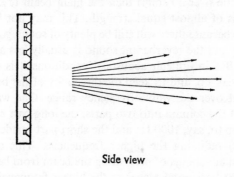

Side view

Fig. 27
Diagrammatic illustration of
loudspeaker column

Plan

Column loudspeakers have two big advantages over ordinary loudspeakers, and indeed over the unaided voice. The first is that they direct most of the speech sound towards the audience. This means that the intensity of the reverberant sound is less, and thus the ratio of direct sound to reverberant sound—necessary for good intelligibility—is much improved. Secondly, the fall-off in intensity of the direct sound with distance can be reduced to some extent, by arranging the column so that the audience further away comes more into the beam.

D*   105

The directivity of a column depends on its length and the frequency of the sound; the longer the column or the higher the frequency the narrower the beam. Thus we cannot get a beam of uniform width over the whole speech frequency range. Another effect is that when the wavelength of the sound becomes comparable with the spacing between individual loudspeakers (for loudspeakers spaced, say 250 mm apart this happens at about 1000 Hz, i.e. when the wavelength is a little over 300 mm) then the main beam breaks up into several beams of almost equal strength. This may not be a serious disadvantage because there will still be plenty of sound going towards the audience, and the reverberant sound is usually less at the higher frequencies. But in difficult acoustical conditions it is desirable to avoid this break-up, and to keep the directivity of the beam reasonably constant over the whole frequency range. One way of doing this is to split the column into two parts, the long part radiating the frequencies up to, say, 1000 Hz and the short part (made up of small loudspeakers) radiating the higher frequencies. This arrangement has the added advantage of preventing the beam from becoming too narrow (in the horizontal plane) at the higher frequencies, which it would do if the bigger loudspeakers were used for the whole range.

Role of the Operator

For any speech-reinforcement system it is advisable to have an operator skilled in the use of his particular system. In addition to the microphone switching, and any loudspeaker switching that may be necessary, his main task will be to control the volume. This is more difficult than it sounds. If only one or two people speak through the system, then the volume can probably be left at some setting found from experience, but in the more usual case of there being a constant succession of different people speaking, their deliveries will vary so much in loudness from one to the other, and from moment to moment, that continuous control is called for. The operator may have a meter to help him, which indicates the loudness of the amplified sound, but most operators prefer to rely on their ears. The greatest fault of operators is that they will have the system on too loud. There appear to be two reasons for this. The first is that the operator may feel that the system is not giving good value for money unless it is on so loud that no one can doubt that it is in use. (The ideal system, on the other hand, should be so unobtrusive that no one realises it is on while still maintaining adequate loudness and intelligibility for everybody.) The second reason is that in any audience there are certain to be a few people who are slightly deaf (perhaps without realising it) and they might have some difficulty in hearing speech that is loud enough for everyone else. One of them may complain to the operator and ask him to turn the volume up. The operator usually complies, thus making it too loud for the rest of the audience. The only solution, and a not very satisfactory one, is to ask such people to sit closer to the loudspeakers. All operators should be warned of these two faults.

It is safer if the system is arranged so that the volume control cannot be turned up so high as to cause feedback. This can be done by using an internal control, when the system is being finally tested, set at 6 dB below feedback level when the main volume control is turned full up.

The operator and his control panel must be placed in the auditorium so that he can see the microphone positions he has to switch, and so that he hears the reinforced speech at about the average loudness for the whole auditorium. If he cannot be put in such a place, then it must be impressed upon him that most of the audience are hearing the speech so much louder or softer than he is, as the case may be.

Frequency Response

All loudspeaker amplifiers are fitted with tone controls, and in unskilled hands they can ruin an otherwise good system. It is desirable for them to be available so that they can be set by the expert when he installs the system and thereafter not touched. When the reverberation time of an auditorium is longer than about two seconds, e.g. in churches, cathedrals and very large auditoria, then the speech is not only more intelligible but also, surprisingly, more natural if the frequency response is limited to between about 250 Hz and 4000 Hz. This restriction must be sharp, i.e. the frequency response must fall off quickly below and above these limits.

Design of Time Delays

Further work has refined the original Haas results, and the effects of several repetitions of the original sounds, of modifying the frequency characteristic of the repetitions, and of the addition in loudness of several repetitions have been investigated. However, for present design purposes we need consider only two graphs. Fig. 28 shows the increase in intensity which a secondary loudspeaker (i.e. a loudspeaker repeating the speech just received from the speaker,

107

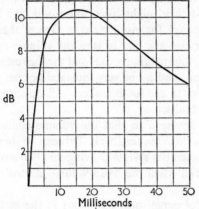

Fig. 28

Increase in intensity (secondary sound over primary sound) for equal loudness as a function of time interval

or a second loudspeaker repeating the speech just received from a first loudspeaker) must have compared with the primary sound if the secondary is to sound as loud. It is plotted as a function of the time interval between the arrival of the two sounds. It is seen that, for example, if the secondary sound arrives at the listener 10 milliseconds after the primary sound, then the secondary sound must be 10 dB greater in intensity than the primary sound if both sound sources are to sound equally loud. In practice, of course, we do not want the two sources (which, for the front part of an audience, will be the speaker and the nearest loudspeaker) to sound equally loud— we want the secondary sound, i.e. the loudspeaker sound, to be not noticeable. Thus the secondary sound must not be as much as 10 dB up (at this 10 millisecond delay), and under laboratory conditions with trained observers listening specifically to detect the secondary sound, it is known that it must not be more than 5 dB up on the primary sound if it is to be completely undetectable. However, in a practical loudspeaker system the audience is not listening so critically; further, there is the visual effect, i.e. the listeners are expecting the sound to come from the speaker, and thus they are predisposed in favour of the sound coming from the 'correct' direction. How much can be allowed for these two factors is not known, but it is probably safe enough when designing a time-delay system to permit the secondary sounds to be 10 dB up on the primary sound over the time-interval range 5 to 25 milliseconds. Of course, if particular conditions do not call for the secondary sound to be as much as this, so much the better.

108

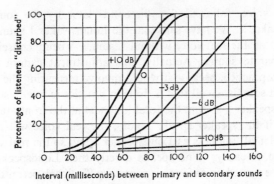

Interval (milliseconds) between primary and secondary sounds

Fig. 29. Percentage disturbance as a function of relative
intensities and time interval

The second graph (Fig. 29) necessary for the design of time-delay systems shows the percentage of listeners who will think the secondary sound is disturbing as a function of the relative intensities of the two sounds and of the time interval between them. For example, if the secondary sound is 10 dB greater than the primary sound, and is 30 milliseconds behind it, then 8% of listeners will be disturbed. It should be explained that the listeners who took part in the original experiments were being very critical and it is probably safe to say that a '10% disturbance' figure would be an adequate criterion. Thus, if the secondary sound is 10 dB greater than the primary sound then it should not be more than 30 milliseconds behind. On the other hand, if the secondary sound is, say, 6 dB less than the primary sound then it can be as much as 80 milliseconds behind before 10% disturbance is reached.

The use of the two graphs can best be illustrated by a trial design. Consider the section of the hall shown in Fig. 30. Loudspeaker 1 is 3 m above the head level of the audience. Listener A will hear first

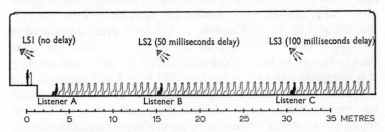

Fig. 30. Longitudinal section of hall with delayed speech-reinforcement system

109

the real voice, followed 5 milliseconds later by the sound from Loudspeaker 1. We have said that the Loudspeaker 1 sound can be 10 dB up on the real voice without Listener A being conscious of it as a separate source. Actually, this adjustment cannot be done simply in practice, because it would involve measuring the loudness of speech sounds, so all that is done in practice is to adjust the gain of Loudspeaker 1 so that the total loudness in the front area sounds correct: if the speaker speaks loudly, the real voice will be of the same order of loudness as Loudspeaker 1 and so the real voice will predominate easily; if he speaks quietly then Loudspeaker 1 will have to be turned up as far as is necessary (feedback permitting), and if this exceeds the permissible 10 dB difference, nothing can be done about it. But even if this state of affairs is reached, the result will still not be too disturbing, because except for those listeners very close to the speaker, he and Loudspeaker 1 will be in practically the same direction.

As we get further away from the speaker, the direct sound and the sound from Loudspeaker 1 will fall off and further reinforcement will become necessary. Let us assume that this happens at 15 m from the speaker; Loudspeaker 2 can now be installed, let us say 3·3 m above the listeners, with an electrical delay in circuit equal to the difference in the times taken for the sounds from Loudspeaker 1 and from Loudspeaker 2 to reach Listener B, plus an extra few milliseconds for the Haas effect. The time taken for the Loudspeaker 1 sound to travel to B is about 16·5/340 seconds (340 m/s being the velocity of sound), i.e. 50 milliseconds, and the time taken for the Loudspeaker 2 to travel is 3·3/340 seconds, i.e. 10 milliseconds. The extra milliseconds to be added depend on how many more loudspeakers are to follow No. 2. If No. 2 is the last, then the extra delay may as well be 15 milliseconds, because Fig 28 shows that there is an optimum effect at 15 milliseconds. If there are further loudspeakers to come, then the extra delay on No. 2 should be less because it will be necessary to fit in the later loudspeakers. Let us make the extra delay on No. 2 equal to 10 milliseconds, so the total delay introduced is $50 - 10 + 10 = 50$ milliseconds. By the time we have got as far back as this in the hall, we can assume that the real sound reaching B is negligible compared with the sound from Loudspeaker 1. This is because Loudspeaker 1 is operating at a much louder level than the man is speaking, and additionally, perhaps, because the real sound will have been attenuated by passing at grazing incidence over the heads of the audience, compared with the sound from Loudspeaker 1 which has a 'clear path' from Loudspeaker 1 to B. The loudness of Loudspeaker 2 can be adjusted so that its sound when it arrives at Listener B is 10 dB up on the sound arriving from Loudspeaker 1.

At this stage it will be useful to consider diagrammatically the arrival of the various sounds at the various positions. Fig. 31 shows

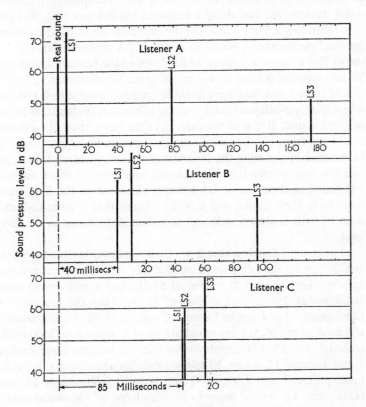

Fig. 31. Arrival of sounds in hall (Fig. 30)

that, at A, the first sound to arrive is the real sound at a level, let us assume, of 63 dB. Next arrives Loudspeaker 1 sound, 5 milliseconds later, and this can be at 73 dB. The total loudness at A will be near enough 73 dB. (We will deal with the sound from Loudspeaker 2 in a moment.) At B, ignoring the real sound, the sequence is: Loudspeaker 1 sound (at a level of 73–10 dB = 63 dB, the −10 dB being due to the distance ratio of 5·2 : 16·5 m) and the Loudspeaker 2 sound

111

adjusted to be 10 dB up on the Loudspeaker 1 sound, i.e. at a level of 63 + 10 dB = 73 dB, and arriving 10 milliseconds later. The loudness at B is thus 73 dB. Now the Loudspeaker 2 sound will also travel back towards Listener A, and if the loudspeaker radiated equally in all directions the loudness of Loudspeaker 2 at A would be 73–12 dB = 61 dB. This is 12 dB down on the Loudspeaker 1 sound at A and it is 73 milliseconds behind the Loudspeaker 1 sound, owing to the electrical delay introduced (50 milliseconds) plus the time taken to get from 2 to A (38 milliseconds) minus the time taken for the Loudspeaker 1 sound to get to A (15 milliseconds). Fig. 29 shows that a secondary sound 12 dB down on the primary sound and 73 milliseconds behind it will cause negligible disturbance, but it should be said here that there might be speech-reinforcement systems where this long-delayed sound could be of comparable loudness to the primary sound. If this is the case, then the amount of disturbance can be got from Fig. 29 and if found to be more than 10% can only be reduced by reducing the loudness of the long-delayed sound. In practice this means that the delayed loudspeakers must radiate less sound to their backs, i.e. towards the front of the hall, than they do to their fronts, and it will be found that a 'front-to-back' ratio of 6 dB is always sufficient to reduce disturbance to the 10% level.

15 m beyond Loudspeaker 2 we shall need another loudspeaker, No. 3. Similar calculations show that Listener C will receive the Loudspeaker 1 sound at a level of 57 dB, and 1 millisecond later Loudspeaker 2 sound at a level of 60 dB. Thus the loudness of Loudspeaker 3 arriving at Listener C can be 70 dB. Loudspeaker 3 will need an electrical delay equal to that of Loudspeaker 2 (50 milliseconds), plus 50 milliseconds for the distance between Loudspeaker 2 and Listener C, minus 10 milliseconds for the distance between Loudspeaker 3 and Listener C and plus 10 milliseconds for the Haas effect, i.e. a total delay to be introduced of 100 milliseconds. The total sequence of sounds arriving at Listener C is shown in Fig. 31.

It should be noted that the loudness at Listener C is 70 dB, compared with that at Listener A of 73 dB. We have thus got 30 m away from the man speaking for a drop in loudness of the apparent direct sound of only 3 dB, and if the system is made 1 dB too loud at A, it will be only 2 dB too quiet at C. These differences would not be noticeable, and we have still maintained the illusion that all the sound is coming from the correct direction. For comparison, if only

Loudspeaker 1 were used, the level at A would still be 73 dB, but at C it would be down by 16 dB.

We have been considering only Listeners A, B and C in our example, but a moment's thought will show that if the system is correct for them it will be correct for the listeners behind them. This is because the relative times of arrival will remain the same, and the amplitude of the first received sound, e.g. Loudspeaker 1 for Listener B, will fall off more slowly as we go towards the back than will the loudness of Loudspeaker 2. Thus the difference between the two loudnesses will decrease, and the Haas effect will still be all right.

We should repeat that the use of time delays not only maintains the illusion, but is a very great help to the intelligibility.

If the same type of loudspeaker is used at all positions—as will often be the case—then the relative levels can be set by adjusting the voltages applied to the loudspeakers. In the above example, Loudspeaker 1 had to produce a level of 73 dB at position A which was 5·2 m away. Thus at a standard distance of say, 1·5 m, Loudspeaker 1 would have to produce a level of $73 + 11$ dB $= 84$ dB, the 11 dB being due to the distance ratio of 5·2 : 1·5. Similarly, Loudspeaker 2 had to produce a level of 73 dB at B, this time at a distance of 3·3 m, so at 1·5 m Loudspeaker 2 would have to produce a level of $73 + 7$ dB $= 80$ dB. Thus when the system is set up originally, some convenient test voltage should be inserted early in the circuit and the voltage appearing across Loudspeaker 2 should be set to be 4 dB lower than the voltage across Loudspeaker 1. They will, of course, be operating from separate power amplifiers because they have different time delays, so this voltage adjustment is easily made. Once all the loudspeakers have been set correctly, there should be no further need to change them; the gain of the whole system should go up or down as a whole, depending on the speaker.

Design of Loudspeaker Columns

The requirements of a good loudspeaker column are (a) that it should be directional in the vertical plane, (b) that this directionality should not vary more than is possible with frequency, (c) that it should be non-directional in the horizontal plane, (d) that its radiation to the back should be at least 6 dB down on the radiation to the front, at all frequencies, and (e) that it should be as small as possible.

The directivity of a column depends on its length and on the frequency of the sound: the greater the length and the higher the frequency the greater the directivity. In detail, the directional characteristic of a source consisting of a number n, of equal point sources radiating in phase, located on a straight line and separated by equal distances d, is given by

$$R_\theta = \frac{\sin\left[(n\pi d/\lambda)\ \sin\theta\right]}{n\sin\left[(\pi d/\lambda)\ \sin\theta\right]}$$

where, at a large fixed distance from the source, R_θ is the ratio of the pressure at an angle θ to the pressure for an angle $\theta = 0$ (the direction $\theta = 0$ is at right angles (normal) to the line), and where λ is the wavelength. In the limiting case where n approaches infinity and d approaches zero so that nd = 1 = the length of the line, we have the ideal straight-line source. The equation then becomes

$$R_\theta = \frac{\sin\left[(\pi l/\lambda)\ \sin\theta\right]}{(\pi l/\lambda)\ \sin\theta}$$

In the practical cases where the source is made up of a number of loudspeakers mounted close together, the simpler formula can be used with sufficient accuracy provided that the distance between the loudspeakers is small compared with the wavelength. The polar diagram (up to 30 degrees either side of the axis) in the vertical plane of a column 3·3 m long at 1000 Hz is shown in Fig. 32. It is seen that there

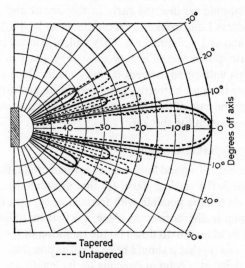

32. Directivity of a 3·3 m column at 1000 Hz

114

are secondary lobes, the greatest of them being 13 dB below the main lobe. If the line source is 'tapered' in strength so that the sound from each element varies linearly from a maximum at the centre to zero at either end, the directionality is given by

$$R_\theta = \frac{\sin^2 [(\pi l/2\lambda)\ \sin\ \theta]}{[(\pi l/2\lambda)\ \sin\ \theta]^2}$$

Fig. 32 also shows the directionality at 1000 Hz of a 3·3 m tapered source, and it is seen that the main lobe is slightly broader while the first of the secondary lobes is reduced to 27 dB below the main lobe. The advantage of this relative suppression of the subsidiary lobes are, first, that less of the column's energy goes away from the audience and therefore there is less reverberant sound, and secondly that there may be some surface in a building, e.g. a dome, which might focus back to the audience one of the subsidiary lobes, thus perhaps causing an echo. As tapering is simply done, either with a tapped transformer or with resistances, and as the power-handling requirements of any column used for speech indoors are so moderate, it seems better to use tapered columns. Other forms of tapering are, of course, possible, e.g. following a binomial law, which suppresses the side lobes even further. When linear tapering is used, the voltages at the ends of the column would in theory be reduced to zero. There is obviously no point in having a loudspeaker at each end of a column with no voltage applied to it, so the tapering in practice is worked out assuming an imaginary loudspeaker at either end. For example, a column consisting of a central loudspeaker with four loudspeakers either side of it would be tapered so that the loudspeaker next to the centre one (and of course the corresponding one on the other side of the centre) would have 80% of the voltage on the centre one, the next 60%, the next 40% and the end ones 20%.

The smaller the individual loudspeakers used in a column and thus the greater number that can be got into a given length, the nearer its behaviour approaches the 'ideal' line source. It has been found in practice that ordinary 150 mm loudspeakers used in a column are indistinguishable in quality from the best 250 mm loudspeakers (at least under reverberant conditions) and therefore the 150 mm size may as well be used. As has been mentioned earlier, the use of the whole length of the column for all frequencies has three undesirable features. The first is that a stage is reached as the frequency increases where the column may well become too directional.

115

There has been an example in a very reverberant building where the beam from such a column was so sharp that, at some positions, speech was perfectly intelligible when the listeners were seated but became unintelligible when the listeners stood up. This was an extreme case, but it does indicate that even in less critical auditoria such sharp focusing is undesirable. The second undesirable feature is that at a still higher frequency, when the spacing between individual loudspeakers becomes comparable with the wavelength (with 150 mm loudspeakers at 200 mm centres this will happen at about 1500 Hz) the column stops behaving as a line source, and the directivity pattern breaks up into a series of lobes of more or less equal amplitude. Thirdly, we want the minimum directionality in the horizontal plane, and even 150 mm loudspeakers will become appreciably directional in this plane at the higher frequencies.

All these defects can be much reduced by splitting the column into two parts: the whole length will be used for the lower part of the frequency range and the shorter length for the higher part. A successful arrangement is to make the whole length of the column using 150 mm loudspeakers, and to cross-over at 1000 Hz to a column of one-quarter the length made of 60 to 75 mm loudspeakers.

A minor disadvantage of loudspeaker columns is that as the beam gets narrower with increasing frequency, so the intensity on the axis increases, i.e. the frequency response (on the axis) is skewed towards the high-frequency end. The effect of this is halved by splitting the column into two, or the electrical circuit could be adjusted to compensate.

All the loudspeakers must be in phase, and it is desirable to test each one, even if they all come from one batch, to ensure that their wiring is consistent. This is simply done by applying a small d.c. voltage to the loudspeaker and seeing or feeling the direction of motion of the coil. Also, all the loudspeakers should have equal acoustical efficiency, and while speakers of the same type do not usually vary much, for complete safety—and to ensure against any defective ones—a simple comparative listening test with a steady signal applied (speech or music) is adequate.

The two parts of a column will probably have different acoustical efficiencies because of the different sizes of loudspeaker used. Usually the larger loudspeakers will be more efficient, and thus either the transformer tappings should be adjusted to compensate for this, or an attenuator should be fitted in the low-frequency part.

Other points about design are: (a) separate pairs of wires should

SPEECH SYSTEMS

be run to each loudspeaker in the column; otherwise (i.e. if a common return is used) the voltage drop along this common return may be sufficient (owing to the comparatively heavy current taken by the centre loudspeakers) to upset the voltages fed to the loudspeakers near the ends of the column; (b) the smaller loudspeakers must be mounted flush with the surface, otherwise their radiation to the sides may be restricted; (c) a resonance may occur behind the loudspeakers between the two parallel sides of the column, and this can be stopped by making these sides non-parallel internally, i.e. by fitting an oblique wood block along the length of one side; (d) the sound from the loudest loudspeakers at the centre of the column should not be allowed to spread internally along the length of the column, and this can be prevented by baffles or by filling most of the vacant space with sound absorbent.

One visual disadvantage of columns is that they nearly always have to be tilted forward (see below) and this often jars with the architecture, particularly in cathedrals where there are strong vertical lines. Columns more sophisticated in design than those described in this book are now becoming available, e.g. in which the main beam is tilted down electrically, but a full discussion of all possible designs of columns is outside the scope of this book.

It is nearly always desirable for a column to radiate less sound towards its back than it does towards its front. At the higher frequencies this will happen if the back is simply boxed in, but at the lower frequencies the sound will diffract round the column. A better design is to use for the back a layer of material which slows up the propagation of sound. Briefly, the effect is that the back radiation from the loudspeakers is delayed by the material by a time equal to the time taken for the front radiation to reach the back round the sides of the column. As the back radiation is, of course, 180 degrees out of phase with the front radiation, the total radiation to the back is reduced by cancellation. With careful selection of the material the front-to-back ratio can be made as much as 20 dB.

Use of Loudspeaker Columns

Columns are made commercially in many sizes and many designs. How long a column is necessary for a given hall cannot be stated with any certainty: it is largely a matter of judgement. A very rough guide would be:

117

Distance to be covered by column (m)	Reverberation time at 500Hz (hall full) (seconds)	Length of column required (m)
30	4 or longer	3·5 to 3
15	4 or longer	3 to 2·5
30	2 to 4	2·5
15	2 to 4	2·5 to 1·7
30	1·5 to 2	1·7
15	1·5 to 2	1·7 to 1·2

The two acoustical factors governing the orientation of loud-speaker columns are, first, that the main beam of sound should fall on the audience area, not only for the obvious reason that this is where the sound is required, but also so that this main sound is absorbed by the audience and does not become unwanted reverberant sound, and secondly, that the directivity of the column should be used to compensate as far as possible for the fall-off in intensity with distance. To some extent these two requirements work against each other, as will be shown.

Consider, for example, a hall with a flat floor 30 m long and employing a 3·3 m linearly-tapered column. This is illustrated in Fig. 33. The column could be placed with its bottom at ear height

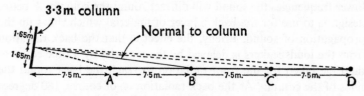

Fig. 33. Effect of distance and directivity

and inclined forward so that the normal from the centre of the column meets the audience at 15 m away, i.e. halfway back. With this arrangement the main beam is falling on the audience area, and we should now consider how much distance compensation will occur. Of course, at the lower frequencies we shall get little compensation because the beam is too wide. But it is the higher frequencies (i.e. above 1000 Hz) which are the most important for intelligibility, because the energy of consonants is in this range, and it is here we can get some useful distance compensation.

118

For comparison we will consider the two frequencies of 250 Hz and 4000 Hz. If we take listener position B (Fig. 33) as our reference point, we can prepare a table of the variation in loudness due to distance and directivity as follows:

Position	250 Hz			4000 Hz		
	Distance factor	Directivity factor	Total (dB)	Distance factor	Directivity factor	Total (dB)
A	+6	−0·5	+5·5	+6	−6	0
B	0	0	0	0	0	0
C	−3·5	0	−3·5	−3·5	−0·5	−4
D	−6	0	−6	−6	−1·5	−7·5

Thus the range of intensities from A to D at 250 Hz is 11·5 dB, i.e. we have gained only 0·5 dB compared with the range of 12 dB a non-directional loudspeaker would give, but at 4000 Hz the range is 7·5 dB, i.e. an improvement of 4·5 dB.

If the column is pointed still further back, then an appreciable amount of the sound energy in the main beam will miss the audience area and will strike the rear wall, after which, if the rear wall is reflecting, it will become unwanted reverberant sound. Further, the front of the listeners might be so far off the axis that, for the highest frequencies, they will be in a null area of the directivity pattern. If the rear wall is absorbent then the first point is not important; if the front listeners get enough sound from the speaker then the second point is not important. In general, the best compromise is to point the loudspeaker column about two-thirds to three-quarters of the way back.

In some cases it might be more convenient to sacrifice the compensation for distance, and mount the column fairly high up (as illustrated in Fig. 34) and pointed so that all the main beam strikes the

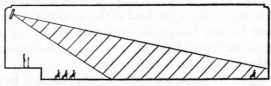

Fig. 34. High column on a flat-floored hall

119

audience. The main advantage of a column—its reduction of reverberant sound—is thus kept, and feedback troubles may be made easier because the microphone is well out of the column's main beam.

If some part of the building, e.g. a pillar, comes between the loudspeaker and the listener then it will throw an acoustic 'shadow'. How serious this is depends on the size of the obstruction, but every effort should be made to avoid having listeners behind any object more than 300 mm wide.

The Use of Speech Systems in some Typical Auditoria

We shall here consider the designs of speech systems for a few typical auditoria, in the light of the above discussions. It is difficult to say what size auditorium will need a speech system although, as already mentioned, they are being introduced into smaller and smaller rooms. A hall used for a variety of purposes, e.g. a school hall, with a flat floor, with some intruding noise and often with indifferent speakers, will probably need one if it holds 200 or more audience. On the other hand, a proscenium theatre with raked floor, with quiet conditions and with trained actors (who manage to speak about 10 dB louder than untrained people without obvious effort), could probably hold up to 1000 audience before a reinforcement system was necessary. There is no doubt that in a new building provision should be made for a speech system if there is any possibility of one being required. Otherwise, there is a danger of temporary systems with bad loudspeaker placing being introduced at a later date.

We should distinguish between two classes of auditoria. The first is where acoustics are good, i.e. the reverberation time is correct for speech and there are no specific defects such as echoes or 'dead' areas; the second is where the acoustics are bad. For the good auditoria, a system will only be needed to make the speech louder, particularly at the back of the auditorium. An example is a hall with a flat floor seating 500 audience with the speaker on the stage. His unaided voice will not reach to the back because of the attenuation due to distance and grazing incidence. A single loudspeaker placed two to three metres above the head of the speaker will make things much better because the grazing incidence will be removed and because it can be operated at a higher level than the man can speak without being too loud for the front rows.

At the higher frequencies a single loudspeaker may be so directional that the intelligibility for those off the axis of the loudspeaker, i.e. in this case those seated near the front and to either side, is less. Further, feedback may be troublesome with the loudspeaker only a metre or two above the microphone, although this can be minimised with a properly orientated directional microphone. The most common method of overcoming these two disadvantages is to have a loudspeaker on either side of the centre line, i.e. either side of the proscenium arch. Thus nobody will be very far off the axis of one or other of the loudspeakers. The loudspeakers will be further from the microphone and feedback will be reduced, but there are two disadvantages. The first is that those seated on the centre line of the hall will receive two loudspeaker sounds. The real sound will arrive first, and if it is loud enough it will determine the apparent direction of the speech as coming from the speaker, which is good. If, however, it is not loud enough, the speech will appear to be coming from either one side or the other, depending which way the listener leans in his seat. This can be irritating. If there is a centre gangway, then the problem does not arise. The second disadvantage is that for those of the audience who are nearer to either of the loudspeakers than they are to the speaker, the speech will appear to be coming from the loudspeaker. On the whole, if feedback can be avoided, the best solution is to have one central loudspeaker unit with a cross-over network and a small loudspeaker to overcome the directivity.

If two loudspeakers either side of a stage are used and if the auditorium is large enough and is fan-shaped or horseshoe-shaped, then the time interval between the arrival of the two lots of loudspeaker sound might be long enough to cause some interference with the speech. For example, consider the plan shown in Fig. 35. The

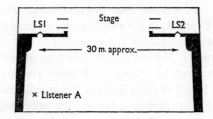

Fig. 35
Plan of theatre with two loudspeakers

sound from Loudspeaker 2 will arrive at listener A about 75 milliseconds after the sound from Loudspeaker 1, and as there is a factor of two in the distances listener A is from the two loudspeakers, the sound from Loudspeaker 2 will be 6 dB down on the Loudspeaker 1 sound. Fig. 29 shows that when the secondary sound is 6 dB down

121

on the primary sound and is 75 milliseconds later, then the disturbance figure is about 10%. Only under severe conditions, e.g. rapid speech in a rather reverberant auditorium, would this disturbance figure be significant, and this example has only been given because such conditions did once occur in the authors' experience, giving rise to severe complaints from some members of the audience. The trouble was stopped by replacing the two loudspeakers with one over the centre of the proscenium arch.

Our second example is a hall with otherwise good acoustics, but with a bad area, such as often occurs underneath a deep balcony. This can be dealt with by placing loudspeakers in the soffit of the balcony, but it is then essential for realism and often for intelligibility that these loudspeakers should be time-delayed. Several examples have occurred where such loudspeakers without time delays have been installed, but were found to do more harm than good. The time delay introduced should, of course, correspond to the time taken for the sound from the stage to reach this area, less the time taken for the sound from the subsidiary speakers to reach this area, plus about 15 milliseconds, and the amplitude of these subsidiary loudspeakers should be such that the sounds from them reaching the listeners should not be more than 10 dB up on the sounds reaching them from the front of the hall.

We have so far been considering only halls whose acoustics are good. When we come to deal with halls with bad acoustics (and this usually means their reverberation time is too long), then there are two solutions. The first is a low-level system, i.e. a large number of small loudspeakers distributed so that every person in the audience is close enough to a loudspeaker for the direct sound—from this nearest loudspeaker—to predominate over the reverberant sound. This is the most foolproof system; in the extreme case each member of the audience has his own loudspeaker and is so close to it that the room acoustics do not matter. However, with this type of system, although the intelligibility will be excellent—if there are enough loudspeakers—there will be no realism because the speech will be coming from the nearest loudspeaker and, further, the large number of loudspeakers all operating at the same level produce a rather unpleasant-sounding effect. To maintain realism—where this is important—it is better in difficult acoustical conditions, to use loudspeaker columns of the appropriate length.

On p. 118 the use of loudspeaker columns to cover a flat area was described. Similar reasonings apply if the auditorium floor is raked, and if there is a balcony, two colums may be used, one to cover the ground area and the other to cover the balcony, as illustrated in Fig. 36. The two columns should be a few metres apart to minimise interaction between them, but should not be so far apart (say, not more than 10 m) as to cause trouble due to the time interval between arrival of the two sounds in areas of the auditorium covered by both columns.

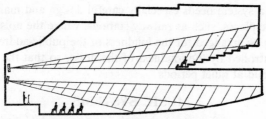

Fig. 36. Use of two columns in a hall with balcony

Probably the worst buildings for speech are large churches and cathedrals. Their reverberation times are long, and the distances to be covered by the loudspeakers are great. Further, the system has to work not only when there is a full congregation, but also when the floor area is only sparsely covered. Loudspeaker columns are invaluable here, but in the largest cathedrals one column will not cover the whole area, and subsidiary, time-delayed columns will also be needed. There are several successful systems in operation, but like all systems they need regular checking and maintenance.

One of the most awkward of all speech-reinforcement systems is when the speech may be from more or less any position in the room. This happens in debating chambers, sometimes in banqueting halls, sometimes in halls used for political or other discussions when the audience participates and in some types of religious buildings where the congregation participates. The usual solution is to use a low-level system with several microphone positions, or a trailing microphone, and usually the loudspeakers closest to the microphone in use are switched off or reduced in volume. With such systems it is very desirable to have an operator in attendance all the time, versed in the use of the system; otherwise some of the microphones not in use are invariably left on by mistake, which reduces the intelligibility and may cause feedback. Some systems have a microphone for every two or three people, and a refinement is to arrange the room into areas, the loudspeakers in each area being delayed appropriately and automatically depending on which microphone is in use.

Arrangements must be made for whoever is in charge in the room, e.g. a chairman, to be in control of the operator so that he, the chairman, rather than the operator, decides who is going to speak next. Or sometimes, if he is used to the system, the chairman can work it himself. Yet another possibility is for the microphones to be voice-operated, i.e. when someone speaks only the microphone closest to him stays in operation, all the others being automatically switched off; such a system needs especially careful design and maintenance.

In some places, such as railway stations, where the noise varies a lot, it can be arranged for the loudness of the public-address system to be controlled by the loudness of the noise, so that the system is not too loud at quiet periods.

MUSIC SYSTEMS

The advice on music systems can only be very general, partly because there is such a variety of circumstances, ranging from pop music calling for kilowatts of loudspeakers, to a harpsichord in a concert hall, and partly because—as already mentioned at the start of this chapter—in auditoria where music systems are important, there will often be a local expert.

Much of what has already been said about speech systems also applies to music systems. First, as for speech, one of the main reasons for the use of a music system is to make the music louder and in fact, as already mentioned, many types of musical performances have developed because of, and rely entirely on, loudspeaker systems. However, unlike speech systems, music systems are not usually designed to mitigate bad acoustics. Secondly, as for speech, good-quality microphones are essential, but their placing will need expert guidance so as to obtain correct balance between, say, a singer and an orchestra. Feedback is not usually so much of a problem because the original source of the sound, e.g. a singer, is usually louder than a speaker, or is more used to holding the microphone close to his or her mouth. Thirdly, most of the remarks about loudspeaker placement for speech systems also apply to music systems, although the space of time in which the repetitions should be concentrated (25 milliseconds for speech) can for music be up to about 75 milliseconds. (One special case of loudspeaker placement is for 'foldback', i.e. loudspeakers positioned fairly close to the performers so that they can hear themselves better.) Fourthly, the advantages that column loudspeakers have for speech systems (e.g. keeping the ratio of direct to reverberant sound high), do not apply to music systems and may, in fact, be a positive disadvantage for music where more reverberant sound is required (see Chapter 4). Further, the directivity of a column varies with frequency and thus the frequency response varies (both on and off the axis), not usually enough to be troublesome for speech, but undesirable for music, and the individual loudspeakers making up a column will not usually be of good enough quality for music. Ordinary good-quality loudspeakers, whether moving-coil or horn, should therefore be used for music systems. (There are exceptions: some column loudspeakers are designed specifically for speech and music.)

One of the few instances where music systems can be used to mitigate difficult conditions is where the building is extremely large, or there are obstructions in it, so that the unaided music does not carry to the remoter parts. An example would be a cathedral where, with the choir at ground level and the organ higher, the organ sound will carry to the west end of the cathedral much better than will the choir. Or the choir may be hidden from the nave by a choir screen, with the organ not hidden. The resulting imbalance between choir and organ can be redressed by placing good-quality loudspeakers fairly high up, which discreetly reinforce the choir only.

ELECTRO-ACOUSTIC CONTROL OF REVERBERATION

As described in Chapter 3, one problem with multi-purpose auditoria is the conflicting requirements for speech and music—a shortish reverberation time for speech and one about twice as long for music. One compromise is to have a reverberation time which is rather long for speech and which is rather short for music, say 1·5 seconds, and to rely on the speech-reinforcement system (if there is one) to overcome the rather reverberant conditions. Another way of dealing with the problem is to have the correct reverberation time for speech and then to lengthen it by mechanical means (as described briefly on p. 60) or by electro-acoustic means. (Electro-acoustic methods of lengthening the reverberation time can also be used in such cases, e.g. large concert halls, where it is difficult to get a long enough reverberation by 'natural' means, perhaps for example because the room volume can not be made large enough.)

One electro-acoustic method which has been used quite extensively is to pick up the music by microphones and feed the signal into some reverberation device (another reverberant room, or springs, or a metal plate) which adds reverberation to the signal, and then to feed this reverberated sound back into loudspeakers in the auditorium. A limitation of this system is that for it to be fully effective it requires the microphones to be close to the performers (not more than 2 m distance); otherwise the feedback round the system limits the amount of power which the loudspeakers can put into the auditorium. (This limitation does not apply if the signal is not to be fed back into the auditorium but is to be used only for recording or broadcasting, but of course in this case the auditorium itself is unaffected.)

Perhaps because of this limitation, other electro-acoustic methods have been developed (some of them patented) for situations where close microphone working is not usually acceptable, e.g. concert halls. A comprehensive and detailed description of all possible systems is beyond the scope of this book, but a brief review of what, at the time of writing (1977) seem some of the most promising systems is given here.

One method, known as 'assisted resonance' (invented by one of the present authors) employs a large number of 'channels', each channel consisting of a microphone-amplifier-loudspeaker, with each microphone and its associated loudspeaker positioned in the auditorium so that they are at points of maximum sound pressure of various room resonances (see p. 37). By controlling the gain of each amplifier, the reverberation time of each room resonance can be increased, and thus the overall reverberation time is increased. The channels must not interact with each other too strongly, and this is usually achieved by putting each microphone into a Helmholtz resonator tuned to the particular frequency of each room resonance.

Another method again uses a large number of channels, but here each channel is not isolated from each other, but all interact to give an increase in reverberation. A third method is to utilise a reverberant space above the ceiling, and microphones in the ceiling pick up this reverberant sound and feed it into the auditorium proper.

6

Principles of Noise and Vibration Control

All problems of noise control involve three parts—the noise source, the recipient of the noise and the path between the two. A wide variety of noise sources will be discussed in forthcoming chapters. The noise path may be simple, as in the case of a machine emitting noise, the noise passing through the air to the operator directly. Or it may be more complex, as in the case of aircraft producing noise, which passes through the air, roof, walls, windows and doors, and the air again, to people indoors. The recipients will normally be people who are involved in a wide variety of activities, which suggests the need for varying degrees of noise control.

In this chapter, we will examine the principles governing the transfer of noise from source to receiver over the normal range of situations commonly found in and around buildings, and the remedies available.

Noise Sources

There are two aspects of noise sources which we must examine. First, the extent, type and manner of noise emission must be known before any meaningful assessment of its effect can be attempted. Second, we must know whether it is practical to reduce the noise output at source. This may reduce the effectiveness of the source, make access difficult or cause the source to get too hot, and so on.

Reduction of noise at source is too large a field to be discussed here. Nevertheless, it is important to remember the possibility, and the value of tackling noise problems at source, by treatment, re-selection, re-siting or timetabling. In many cases, the large cost of noise-control constructions and treatments can be reduced substantially by tackling the source. For the moment we will assume that we are unable to alter the source and examine the noise emission.

A source may emit sound energy equally in all directions (this is termed an omnidirectional source). Usually, it is directional, i.e. more energy is emitted in certain directions than others. The spectrum shape and level of the noise will vary a great deal from source to source. It may be stationary or moving. It may be part of a larger source, such as a vehicle in a stream of traffic. It may be intermittent or continuous.

It is often possible to obtain information about the total noise output of a source. But if the output is not even, we must know how much is radiated in the direction which interests us. Fig. 37 shows

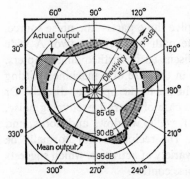

Fig. 37

Example of variation in noise output from a machine with direction (in a horizontal plane)

the directional nature of a machine in terms of a plot of sound pressure. If twice the average energy is radiated in one direction, the source directivity will involve a factor of 2, resulting in an increase of 3 dB in sound pressure level in that direction. There is a tendency for larger noise sources to be more directional owing to mutual reinforcement by different parts of the source. Very close to the source, in what is termed the 'near field', levels may vary substantially as a result of sound waves from different parts of the source cancelling or reinforcing one another.

NOISE OUT OF DOORS

The geometry of the source not only affects directivity, but also the reduction in sound pressure level with distance from the source. If we assume 'free field' conditions, i.e. the absence of reflecting surfaces, from the inverse square law (see p. 19), we expect a 6 dB reduction with doubling of the distance from a small source. But with a 'line' source, (e.g. a busy lane on a motorway) or large sources, (e.g. very large machinery), the reduction in level with distance is smaller, until we are far enough away from the source for it to behave like a point source—this is usually about one-third of the largest dimension of the source (see Fig. 38).

But, in practice, there are normally reflecting surfaces to consider,

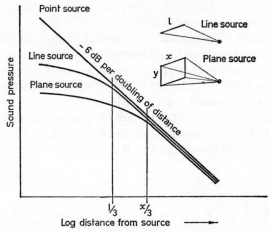

Fig. 38. Noise levels near sources of different shape

even out of doors. The ground is perhaps the most important (see Fig. 39), although in built-up areas, the effects of reflection from and

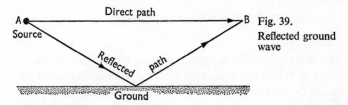

Fig. 39.
Reflected ground
wave

between buildings can be substantial, e.g. the 'canyon' effect (see Fig. 40). Attenuation of noise by the ground varies with the type of

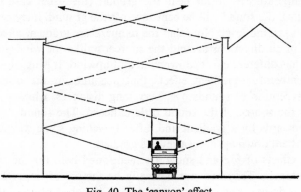

Fig. 40. The 'canyon' effect

E

cover—concrete, grass, crops, snow, etc., and the relative positions of source and receiver (see p. 217).

The air itself will attenuate noise at high frequencies. The effect usually becomes important only from 1000 Hz upwards. Absorption by the air changes with temperature and humidity. Detailed values are given in Table VII, Chapter 9. The weather affects sound propagation. Where there is a wind, there is normally a wind gradient. In Fig. 41 the wind at the bottom is not blowing as fast as it is at the top. The difference in speeds produces a bending of the sound upwards, resulting in less energy near B than there would be otherwise.

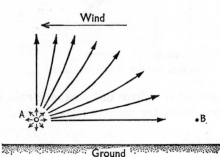

Fig. 41

Effect of wind gradient on the propagation of sound

Temperature gradients have a similar effect. This is because the velocity of sound increases with increase in temperature. If the temperature of the air is higher near the ground than it is in the upper layers (the usual case during the day), the sound waves higher above the ground will travel slower and the sound will be bent upwards, resulting in quieter conditions at ground level. Conversely, when the temperature is lower near the ground (the usual case during the night), the sound will be bent towards the ground, increasing the noise at ground level. Note that the temperature gradient affects the sound in all directions around the source, unlike the wind gradient which has different effects downwind and upwind. It is usually wind- and temperature-gradient effects that account for the occasional freak reception of sounds over very long distances, while at places nearer the source of the sound little is heard. The sound has been bent upwards by a gradient and, after travelling some way at high level is bent down again by a reverse gradient.

The effects of fog and snow are mentioned here for the sake of completeness. They will not normally be factors of importance in building design. Fog causes an increase in the absorption in the air.

A moderately dense fog (visibility 50 metres) gives extra attenuation of 1 to 3 dB per 100 metres, depending on frequency. Snow forms an absorbent layer on the ground and this affects the reflection at the ground, thereby reducing the sound level at a distance.

Weather conditions can therefore produce substantial changes in noise propagation, making it extremely difficult to predict external noise conditions accurately over any distance. Variations of the order of $\pm$ 10 dB are common. This, of course, presents a problem when decisions have to be taken on the effect of this noise on buildings. In practice, it is usually reasonable to make some allowance for adverse wind and temperature gradients but catering for the worst case is likely to be an overdesign. The effects of fog and snow are normally ignored in the United Kingdom. The weather can also become a significant source of noise in buildings. In particular, wind and rain are sufficiently common to be of concern. The passage of wind around buildings, into ventilation grilles and past other external elements can result in noise which is very disturbing. The impact of rain on lightweight roofing or roof lights can produce high internal noise levels.

There are three main approaches to the control of noise out of doors, apart from control at source. First, we can plan to allow as great a distance as possible between the source and the recipient, with 'soft' landscaping provided between. Second, we can place a screen between the two. Third, we can rely on the construction of buildings to resist noise penetration. The value of separating source and recipient by distance is self-evident, but, it must be remembered, with reference to the inverse square law, the distance required to achieve significant reduction in noise level increases as we move away from the source. Usually the freedom to plan in this way is restricted and screening or sound insulation are necessary.

If a barrier or screen is placed in the path of noise propagation (e.g. a solid fence, a wall, an earth bank or a building), the attenuation achieved increases towards the higher frequencies. At high frequencies, screens produce an acoustic shadow, similar to that produced by light. But, as the wavelength increases towards the lower frequencies, the sound bends further round the edge of the screen, filling the shadow zone (see Fig. 42). The benefit of a screen is determined by the increased distance which the sound is forced to cover between the source and the recipient. As a result for a given height of screen, the closer to the source or the recipient, the better (see

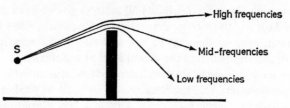

Fig. 42. Screening at different sound frequencies

Fig. 43). The width of the screen is also important; the screening offered by a tall, slim building may be determined by its width rather than its height. Even when a noise source can just be seen over the top of a screen wall, there is a small amount of diffraction around the screen, resulting in attenuation up to 5 dB. Hedges, trees and open fences provide little real noise screening although they may provide visual screening. Calculation procedures for screening are given on p. 218.

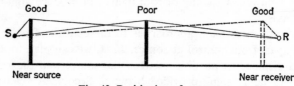

Fig. 43. Positioning of screens

The control of external noise by sound insulation of the building construction is based on principles discussed later in this chapter.

NOISE WITHIN ROOMS

When noise is created within rooms, the noise level at any position may be regarded as being made up of two parts; the first part is the direct sound, i.e. sound travelling directly from the source to the recipient, and the second part is the reverberant sound, i.e. the sound reaching the recipient after reflection from the room surfaces. This is illustrated in Fig. 44. The intensity of the direct sound will fall off as the distance from the source increases (for a small source, by 6 dB every time the distance is doubled). On the other hand, the intensity of the reverberant sound will usually be more or less even throughout the room. Close to the noise source, the direct sound will have the greater influence on noise levels. At positions more remote from the source, the reverberant field becomes more important. The

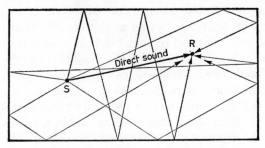

Fig. 44. Direct and reflected sound paths in a room

change in noise level with distance is illustrated in Fig. 45. Proposals for noise control treatment will depend on which of the two components is more dominant.

If we are concerned with noise transmission from one room to another, unless the source is very close to the room boundary, it will normally be the reverberant sound which is of interest. If it is a question of protecting people within rooms, either the direct or the reverberant sound may be important depending on the distance from the source.

The relative levels of reverberant sound can be calculated if sufficient is known about the noise source and the room acoustics. Fig. 45 gives a guide to the likely variation in levels with distance for a range of room conditions. It is however important to note that in long, low rooms with an absorbent ceiling and/or floor, the reverberant sound

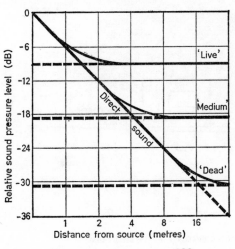

Fig. 45

Change in noise level with distance from a source for different room conditions

133

is not uniform and noise generated at one end of the room will be reduced in level the further we get away from the source. In open-plan areas, with floor and ceiling absorbent, a reduction in reverberant level of 3–5 dB per doubling of distance is to be expected. As a rule of thumb, direct sound will usually be most important within a metre of the source. In practice, this means that anyone operating a machine is likely to be in an area where direct sound predominates.

Control of direct sound from a source within a room cannot be achieved by doing anything to the room boundaries. This must be done at source by enclosure or screening.

Partial Enclosure or Screening

If we place an enclosure around the noise source, the noise level will build up within it to an extent dependent on the size of the enclosure and on its internal absorption. But there will normally be a greater reduction of the noise as it passes through the enclosure. This can be a very useful noise control procedure, but more often than not, there are practical requirements, e.g. to allow access, ventilation or movement of the source, which restrict this method. It is more frequently possible to achieve partial enclosure or screening, although this results in much less noise reduction than complete enclosure will give (see Fig. 46).

If a barrier is placed between the source and the listener (see Fig.

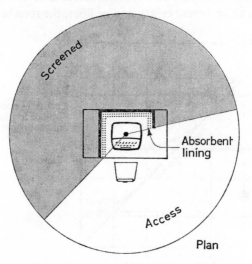

Fig. 46. Local control by partial enclosure

134

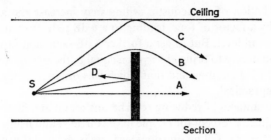

Fig. 47. Indoor screening

47), some of the direct sound is reflected back to the source, some absorbed by the barrier, some transmitted through it and some passes round the edges. Reflection of sound from nearby room surfaces or furnishings may well bypass the screening.

For internal screening, a barrier will not normally need to be very heavy to ensure that the sound passing straight through it is negligible, compared with the amount which is diffracted round the edges or reflected by room boundaries (Paths B and C). The most effective screening or partial enclosure occurs when diversion of the sound around the screen is pushed to a maximum and the nearby room boundaries are highly absorbent. Therefore, to be effective, as with external screening, barriers should be as close to the source and/or the listener as possible and particular attention should be paid to the areas of room boundary surface which could give bypass reflection. The use of absorbent material on the surface of the screen has little effect on screening. But it will limit reflection back to the source and add to the total absorption in the room.

Control of Reverberant Sound

Reduction of reflected sound can be considered in two parts—control local to the source and control by general room absorption. Often, sound which would have been reflected by room boundaries can be collected and absorbed, close to the source, before this can happen. Or the places where reflection would occur may be made absorbent. If the noise gets beyond these points, we can consider the general absorption of multiple reflections (or reverberation) in the normal way. The reverberant sound intensity will be determined by the amount of absorption present in the room. We need to double the amount of absorption present in the room to get a reduction in the reverberant sound level of 3 dB. If a room is originally rather bare,

135

the installation of an acoustic ceiling may increase the absorption present by a factor of four, leading to a 6 dB reduction in the reverberant sound level. But to get a further 3 dB reduction further treatment would have to bring absorption equal to the sum of that already provided by the ceiling and that which was originally present. Often this is impractical.

The advantages of reducing reverberant sound are, firstly, that the room is quieter (except for people in the direct sound field), secondly, the sound energy falling on room surfaces is less and this means that less sound will be transmitted to other rooms. Thirdly, machine operators will hear their own machine, and any verbal warnings of danger, more clearly, if noise from other sources is less. Fourthly, transient noises will die away more quickly, lessening the impression of general clangour.

One way of estimating the value of a material for mid-frequency broad band noise reduction by absorption is to refer to its Noise Reduction Coefficient (NRC)—the mean of absorption coefficients over the four octaves centred on 250, 500, 1000 and 2000 Hz. (See also Appendix A.)

SOUND INSULATION

We now come to the control of noise getting from one room to another, i.e. the problems of sound insulation. We have already said that extra absorption in the room where the noise is will reduce the intensity of the reverberant sound, and thus help the insulation. But the reduction in intensity that this achieves is small compared with the potential reduction by sound insulation. In other words, extra sound absorption is no substitute for adequate sound insulation.

Of the sound which strikes a room boundary, part is reflected, part absorbed within the material, and part transmitted to the other side or to elsewhere in the building. It is most important to avoid confusion between sound *absorption* and sound *insulation*. An example may help in this. Material which offers good sound absorption often fails to provide very good sound insulation. An open window is a good absorber from the point of view of the room in which it is located—but it offers little insulation. Fig. 48 shows a material which offers good sound absorption and converts 75% of the sound energy which it receives into other forms of energy, reflects 5% and allows 20% to pass through. Therefore, although from the source side, 95% of the sound has been absorbed, the transmitted energy has been reduced to one-fifth—in decibel terms, a reduction of 7 dB. This is

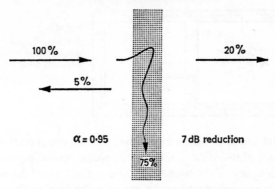

100%

5%

$\alpha = 0.95$

20%

7 dB reduction

75%

Fig. 48. Relationship between absorption and insulation
—example

poor sound insulation. The point is stressed here because it is such a common misconception that sound absorption is the key to solving all noise problems.

We will refer to the room in which the noise originates as the source room, and to the room to which the sound is transmitted as the receiving room. There are two types of sound insulation to be considered: airborne sound insulation and impact sound insulation. The first type concerns insulation against noises originating in the air, e.g. voices. The second type concerns impact noises, e.g. footsteps. This second type of noise is really a combination of airborne and impact noise, because the impacts will produce airborne noise in the source room and this airborne noise will also be transmitted. But in most cases, the noise produced in the receiving room by the direct impact noise predominates.

Further, vibration generated in one room may set the surfaces of a remote room into vibration and they will thus radiate some noise. This is discussed further on p. 150.

We will deal first with airborne sound insulation. A noise source produces a reverberant sound field which impinges on all the room surfaces. If the source is close to the boundary in question, direct sound may predominate. But this is seldom of major significance. The simplest case is when the receiving room is separated from the source room by a single, solid wall with no openings of any kind. The first and most obvious way for the sound to be transmitted from one room to the other is shown as path A in Fig. 49. The sound waves falling on the source room side of this wall cause a bending motion (of course, by very small amounts, which we would not expect to see

E*   137

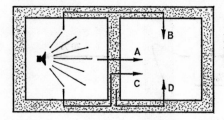

Fig. 49
Paths for sound
transmission between
adjacent rooms

or feel) and sound is radiated in to the receiving room on the far side. The amount of radiation and hence the sound insulation will depend on the frequency of the sound, on the construction of the wall, and, perhaps most important, on its weight. Practically all walls and floors have less insulation at the low frequencies than at the high frequencies.

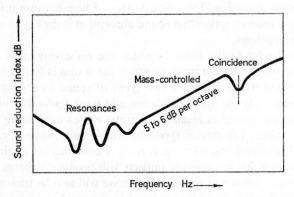

Fig. 50. Typical change in sound insulation with frequency for single partitions

Fig. 50 shows the relationship between sound insulation performance (or Sound Reduction Index—see p. 232) and frequency of sound for a typical single-leaf solid wall. At low frequencies, the stiffness of the construction and primary resonances have the greatest effect on performance. But, as the frequency increases, we move into the zone where the mass of the partition has the greatest effect. Since, for many practical building constructions this zone falls in the middle of the audio-frequency range, the surface mass of a partition is perhaps the most important single influence on the practical insulation provided by it, assuming it is well sealed at joints and edges, and bypass routes are adequately blocked. We also find a sharp dip in performance which occurs when the projected length of the sound waves in the air coincides with the natural flexural waves in the partition (see Fig. 51). This is referred to as the coincidence effect.

In the mass-controlled region, the sound insulation increases by about 5 dB for every doubling of the frequency, i.e. every octave. For example, a single wall with insulation of 30 dB at 250 Hz will have an insulation of approximately 35 dB at 500 Hz, 40 dB at 1000 Hz and so on. However, for convenience, the average insulation over some frequency range is often used.

Whereas this offers a convenient means of classifying performance, it can be dangerous to oversimplify. First, the average value depends on the frequency range selected; the range covered by the sixteen third-octaves from 100 Hz to 3150 Hz is currently recognised widely. But other ranges, such as the 'speech range' (variously

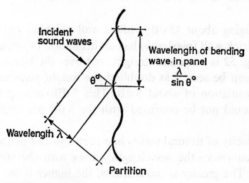

Fig. 51. Coincidence

e.g. 400 Hz–2000 Hz, 250 Hz–4000 Hz) are often referred to—they give a higher average. Second, it is easy to forget that a partition giving a 35 dB average will usually provide very much less insulation at low frequencies. It is obvious that where a single figure is quoted, the frequency range it refers to should always be given. Values of Mean Sound Reduction Index must not be confused with Sound Transmission Class (STC)—a single-figure rating used in the USA in particular, based on the relationship of the performance spectrum to standard curves.

For single walls, the average insulation is strongly influenced by its weight per unit surface area, as indicated by the so-called Mass Law. Fig. 52 shows this law plotted over the range of superficial weights met in practice. It will be seen for example that a wall which weighs 250 kg/m², e.g. a 110 mm brick wall (plastered) has an insula-

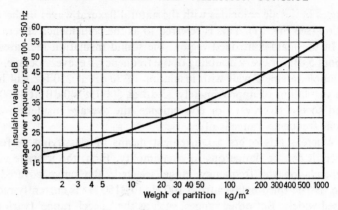

Fig. 52. The 'Mass Law' for sound insulation

tion averaging about 45 dB, while a wall weighing 490 kg/m², e.g. a
230 mm brick wall (plastered) has an insulation of about 50 dB. The
line in Fig. 52 is not quite straight, but over the heavier part of the
range, it can be seen that doubling the weight gives an increase in
average insulation of about 5 dB. This 5 dB relating to change in
weight should not be confused with the 5 dB per octave described
earlier.

The velocity of flexural waves in a partition and hence—for a particular frequency—the wavelength varies with the stiffness of the
partition. The greater is the stiffness, the higher is the velocity and
the longer the wavelength. Thus, for heavy, stiff partitions the coincidence frequency at normal incidence will be low. For example, for
a 230 mm brick wall, it is approximately 80 Hz. For lighter partitions,
the coincidence effect may come at, say, 2000 Hz, which subjectively
is most important. To prevent the loss of insulation at this frequency,
the partition can be made less stiff (but without reducing its weight
appreciably) so as to move the coincidence frequency up to, say,
4000 Hz which will not usually be so important subjectively.

Many partitions used in buildings are not single. An improvement
in sound insulation at higher frequencies can be achieved by making
use of an air space between two skins. For example, if a plastered
230 mm brick wall (50 dB nominal) is split in two, and a 50 mm air
space introduced (wall ties being resilient, e.g. 'butterfly' wire ties),
a small improvement in performance is achieved for the same surface
weight, i.e. a plastered 280 mm cavity wall gives 52 dB nominal. For
such heavy partitions, the effect of the air space is quite small and

140

flanking transmission will limit improvements. But for lighter materials such as plywood, sheet metal or plasterboard, an air space frequently provides up to 10 dB improvement over the equivalent weight of single partitioning.

The typical performance of a double construction is different from and more complex than that shown in Fig. 50. Insulation tends to increase at more than 5 dB per octave. Weakness due to coincidence effects can be reduced if different weights per unit surface area are used in each skin. Additional resonance also occurs as a result of the mass-spring-mass section (the air acting as a spring). In practice, performance of double constructions is best derived from laboratory or field tests of the construction.

We have so far been considering only Path A in Fig. 49. But as the resistance to Path A increases, from about 35 dB upwards, other paths for the sound become increasingly important. The other paths are: Path B, which is due to sound falling on the surfaces of the source room other than the dividing wall, the resultant vibration travelling along in the walls or floor into the receiving room and being radiated there; Path C, which is due to the vibrations of these other surfaces in the source room causing the dividing wall to vibrate (because they are joined to it) in addition to the direct excitation of the dividing wall; and Path D, which is due to vibration in the dividing wall causing the other surfaces in the receiving room to vibrate. All these three paths, B, C and D, along with other bypasses (e.g. open windows in the side walls of the adjacent rooms) are lumped together under the heading of 'indirect' or 'flanking' transmission. How important this indirect transmission is depends on the construction of the dividing wall, the flanking constructions and the way they are connected.

When the sound energy in the structure arrives at a junction such as that in Fig. 53, it will tend to be partially reflected and the remainder will split up and follow different paths. If we insert a structural

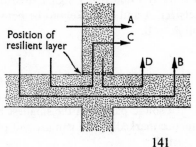

Fig. 53

Sound paths at junctions in the structure

break on one of these paths, the energy is therefore likely to become stronger along the remaining paths. The use of such breaks is only valuable where the diversion of the sound via other paths will reduce the noise transferred to the receiving room of interest. Often, about the same amount of energy arrives at each side of the break (i.e. Paths C and D), so that the rerouting of energy is little different from the case where the structure is not broken.

In Fig. 53, if Path C is impeded, Path B is strengthened. But if Path D is impeded, Path B is reduced. Similarly, Path A is strengthened or reduced by the effects of C and D. Therefore, the overall effect of making a break is often negligible. But there are cases where this type of break is useful (e.g. at the edge of a floated construction —see Fig. 105). It must be properly designed to be effective. The use of a thin resilient membrane is unlikely to effect any significant break, particularly for low- and middle-frequency sound. The material and its geometry must be carefully selected. As a guide, it is useful to bear in mind that low-frequency energy in the structure is best impeded by changes in section, whereas high frequencies are more susceptible to elastic separation (provided the material is not too stiff, as may well be the case under loading). It is a common belief that any resilient material, e.g. natural rubbers or neoprene, will provide a break in a structure, an anti-vibration mounting, an impact-reducing floor covering or a door seal. The properties of materials to be used for these purposes must be carefully selected. They rarely offer low-frequency isolation unless used in shear or large blocks.

It is sometimes necessary to consider noise transmission from one room to another not adjacent to it. In such cases, it is usually flanking transmission that determines the insulation, unless partitions are extremely light. Since flanking transmission is subject to a wide range of variables, it is not possible to provide a simple accurate prediction. However, for a working rule of thumb, the insulation between A and C in Fig. 54 is likely to be 10 dB higher than from A to B, and the insulation between A and D is likely to be between 5 and 10 dB higher than from A to B.

Weakness in Insulation

For a construction to provide its best insulation, it is important that it is free from cracks, holes or other air paths between the source and receiving rooms. An air path is a sound path, and the smallest hole can reduce insulation performance markedly. The gap around pipe-

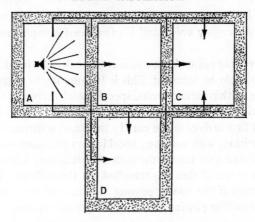

Fig. 54. Insulation between rooms not adjacent to
one another

work passing through a partition, a door ajar, ill-fitting joints—all contribute to a reduction in sound insulation. Porous constructions, such as clinker concrete blockwork or brickwork with open mortar joints, will not perform up to the Mass Law prediction for the same reason. The insulation can be restored by sealing one side—or better, both sides—of the wall, e.g. by plastering.

A most important aspect of sound insulation, often overlooked, is that the total sound insulation of a composite construction is determined to a large extent by its weakest link. A window (20 dB nominal) in a 110 mm brick wall, plastered, (45 dB) is likely to reduce the insulation of the wall to about 25 dB. A detailed method for estimating insulation of composite constructions is given on p. 235. A more extreme example of the loss of insulation due to a weakness is a 25 mm square hole in a plastered 230 mm brick wall, 2·5 metres high and 2·5 metres long. The potential 50 dB average insulation of the brick wall will be reduced to a real value of approximately 40 dB. This emphasises the great importance of sealing all air cracks, if good sound insulation is to be achieved.

Doors are often a weak link. Their superficial weight is usually less than that of the wall they are built in to, and the aircracks round them offer an easy passage for sound unless they are sealed. Another common weakness is the flanking of partitions via a continuous ceiling void running over the top. With perforated or light-weight suspended ceilings, the sound is able to pass through the ceiling, along the void and down into the adjacent room. A similar weakness is via common ventilation ducting serving nearby rooms, although the ceiling void itself is often used as a plenum for ventilation.

An interesting point about this sort of weak link is that it is difficult to detect simply by listening. This is because of the Haas effect (see p. 102), i.e. within certain limits, speech which arrives at the ear from two directions appears to be coming from the direction from which the sound first arrives at the ear. In this case, a listener on one side of the partition, with someone speaking on the other side, will hear first the sound that comes through the partition, followed, a little later, by the sound that has travelled via the ceiling/s. This will be the case even if the sound passing via the ceilings is up to 10 dB stronger than that passing directly through the partition.

Protection by Insulation

The extent to which sound generated in the source room is heard in the receiving room is by no means solely determined by the sound insulation performance (or Sound Reduction Index) of the partition and its flanking paths. For a complete picture, a number of other aspects have to be considered. The sound pressure level difference is also influenced by the surface area of the partition common between the two rooms (the larger the area, the more energy is fed into the receiving room) and the amount of acoustic absorption in the receiving room. If the boundaries and contents of the receiving room are very hard, the sound pressure level in that room will build up—the difference in level between source and receiving room will then be reduced. Conversely, if the receiving room is highly absorbent, the sound pressure level difference between rooms will increase, even though the partition has not changed. Therefore, when specifying performance, we must be clear whether we mean Sound Reduction Index (a property of the partition construction) or Sound Pressure Level Difference (the actual difference in level between rooms). Finally, we must bear in mind that the sound from the adjacent room may not be heard very well because of the presence of other sound, and this sound must also be subject to control.

Impact Noise

Control of noise created by impacts, e.g. footsteps on floors, is important in many buildings. Instead of a partition being excited by airborne sound from a source room, in this case, the partition is directly excited by impacts. A partition which has a very good airborne sound insulation performance, e.g. 250 mm concrete, can provide very little impact noise control.

By far the best means of control is to reduce the amount of impact energy getting into the floor itself, for example using a resilient surface, such as foam-backed vinyl, ribbed rubber or carpet. Such surface layers are not always practicable (and incidentally, do not provide any significant extra airborne sound insulation). They also tend to be of benefit only for middle and high frequency isolation.

An alternative approach is to use a floating floor construction, e.g. a concrete or timber raft laid on carefully selected resilient material on top of the main structural floor (see Fig. 103). The use of glass-fibre or mineral-wool quilt as the resilient material has been widespread in the past and has provided significant isolation in a relatively convenient manner. But more effective isolation can be achieved with mounts which have carefully selected frequency characteristics and loading range. A well designed floating floor can provide substantial improvement in airborne sound insulation as well as impact sound insulation.

With impact sound, the radiation into the receiving room, the effect of absorption in the receiving room, size of the partition, flanking via the structure and other noise are similar to the cases we have described for airborne sound.

Noise from Mechanical Plant

Modern building incorporates increasing numbers of mechanical systems for heating, ventilation, cooling, power generation, lifts, escalators, refuse-disposal units and many other purposes. These systems bring with them more and more noise, which must be controlled sufficiently to allow reasonable conditions for the occupants and to avoid disturbing neighbouring properties.

The largest contribution to such noise control can be made by proper planning of mechanical systems at a sufficiently early stage in the design programme. It is common sense to plan separation of noisy plant such as chillers, large fans or standby generators from locations requiring quiet conditions. Equally, noisy plant should be kept as far as possible from adjacent properties which must not be disturbed, e.g. residential property. Much expenditure on noise control can be saved if sufficient forethought is given to such planning. Of course, there are many circumstances where this separation is not practical. The cost of longer pipe or duct runs or the wider effects on the planning of the building will often restrict the potential for such simple means of noise control.

Often, where mechanical sources are close to or serving critical areas, insufficient space has been allowed to incorporate the necessary noise-control equipment in e.g. plant spaces, builder's work-shafts and ceiling voids. Again, this is a matter of planning.

In general terms, mechanical systems involve a main source of sound energy (e.g. a fan, a pump, a boiler) and linked systems which distribute the sound (e.g. ductwork, pipework, flues). Noise will also be generated within the distribution system, e.g. from air turbulence

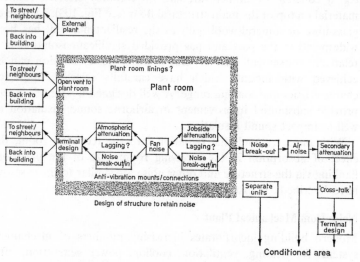

Fig. 55. Aspects of noise control for a typical ventilation system

or cavitation in ducts or pipes respectively. There are also smaller unit noise sources, such as fan-coil units, heat-pump and induction units, although these are often linked to central services.

As an example, Fig. 55 shows the main areas where noise control of a typical ventilation system is necessary. The noise generated by fans is broadly related to the type (axial, centrifugal, paddle-blade), the duty (air volume and working pressure) and the speed of running—see p. 243. It is important that fans, or for that matter any plant, are selected to cope with working conditions. For example, there are a number of noise problems which arise because fans are working at too high a pressure and going into a part-stall condition, which results in excessive noise generation. Nowadays, it is normal to obtain sound power output data (see p. 243) from manufacturers of fans, for given working conditions. We must be concerned with the noise which is radiated from the casing or open inlet of the fan, the noise passing down and/or generated in ductwork and the transfer of vibration into the structure.

Noise radiated directly from the fan or its casing into the plant room, if excessive, must be controlled by the sound insulation of the plant room construction or by local enclosure. The worst problems of direct noise tend to arise with high-duty open-inlet fans. Local enclosure can prove difficult because of the requirement to feed air in and out and to obtain access. Fans are often built into packaged air-handling units, incorporating filters, heater batteries, etc. The fan may then sit in a plenum surrounded by the unit casing. Occasionally, for very high-volume, high-pressure fans, it is necessary to increase the sound insulation provided by these casings and also to consider the sound which is radiated on to the unit base, which may need to be floated.

In general, noise will travel through ducts very efficiently (as sound travels through a speaking tube). The direction of air flow will generally have no significant effect on the transfer of noise, since the speed of sound (approx. 340 m/s) is very high relative to normal air velocities (1·5 to 30 m/s). Therefore, for example, on an air supply system, we must be concerned with both the noise transferred to the conditioned room and the noise travelling against the airstream coming in from the air intake louvres. Part of the noise which passes down the ducts will break out of the duct walls, particularly if they are not very stiff. Very floppy rectangular ductwork is easily moved by the excitation from the sound inside and will allow low frequency noise to break out, whereas stiff, e.g. spiral-wound circular ductwork will retain the noise. The attenuation of noise with distance inside the duct is therefore greater with rectangular ducts, since a lot of energy is breaking out. But if the duct is passing over a critical area the escape of noise from the duct can be unacceptable, particularly if there is only an open or lightweight ceiling between the duct and the listener.

Apart from the loss of energy from the duct in this way, attenuation of duct-borne noise occurs either because of the duct geometry or by insertion of porous absorptive material. Noise is attenuated at bends, branches, T-pieces and at terminations by reflection, screening, absorption and the subdivision of the sound into the branches of the system. This 'natural' attenuation of the system may

147

then be increased, particularly at middle and high frequencies by insertion of absorbent lining of duct walls or the use of pods or splitters, i.e. units of absorptive material set in the airways. Low-frequency noise is attenuated by the use of very thick linings, and by reflection at changes in section or small terminal apertures. Plenum chambers and terminal design can therefore be used to advantage. Vibration transmission paths are equally important and are discussed shortly (see p. 150).

Noise generated within ducts by air movement becomes increasingly important as higher air velocities are used. If air flow is disturbed by an obstacle, e.g. a damper or a sharp edge, turbulence or 'eddies' produced will store pressure differentials, which will be released as noise when they come into contact with ductwork materials or contents. For example, if a damper is almost closed, the air velocity passing the edges tends to be high. If the damper is just behind a supply grille, noise will be likely to occur at the grille (an awkward problem to solve in practice). Generally, this type of noise tends to occur more frequently at the middle and high frequencies, although lower frequencies may be generated where large items react with the air flow. In principle, such noise can be kept to a minimum by maintaining good aerodynamic flow and correct siting of volume-control devices. Nevertheless, current practice renders this type of noise very common.

Fluid systems exhibit similar principles in outline. Excessive airborne noise direct from pumps is less frequent than from fans, although pump motors can often be noisy. Enclosure or screening is appropriate in such cases. Commonly, noise is distributed by or created within the pipework, fed into the structure and re-radiated from the structure which is acting as a sounding board. Since pipes are generally circular and therefore quite stiff, pump noise (impeller, mechanical or motor noise) can travel long distances with little attenuation. Noise is often caused by the actual flow of the liquid through the pipe, e.g. turbulence due to poor fluid flow conditions or 'hammer', which can arise when flow is stopped rapidly by valving off, resulting in multiple reflection of impulses which gradually dissipate the stored kinetic energy. Attenuation in fluid systems is usually more by change of section or side-cavity devices. From the smallest domestic system upwards, this is an area where there is room for substantial improvement in design with noise in mind.

Boilers tend to be noisy at burners and associated induced draught fans. Airborne noise can often be reduced by local burner hoods.

But noise will also be transmitted into flues which discharge the energy and may well impose their own tonal weighting on the noise output. Attenuators capable of withstanding hot gas flow (incorporating changes of section similar to the traditional car silencer—often referred to as reactive attenuators) are often applicable.

Chillers, motor generators, heat-pump units, lift motors, refuse-disposal units, dishwashers and escalators are other examples of noise sources which must be considered. Each must be investigated for its airborne noise output, so that the room in which it is housed can be sufficiently well insulated to protect critical areas. Or sufficient screening or enclosure must be achieved. Attention must be paid to what is attached to it, e.g. its fixing to the structure, associated pipework, mechanical linkages or electrical service connections. Each system tends to have its own characteristic problems, e.g. the chatter of counterweights on guide rails in lift shafts. Space does not allow a full review of all mechanical systems. However, by breaking down the noise problem into the groups—sources, paths and receiving positions, a logical approach to the noise control can be achieved. Where the systems are complex, it is wise to seek expert assistance.

The introduction of mechanical services systems can bring other noise problems which are not directly related to their prime function. A good example of this is 'cross-talk' i.e. the passage of other sound such as speech via ductwork or piping. If an extract ventilation system involves ductwork which passes directly from one room to another, the ductwork will often allow conversation to be heard in the adjacent room because of this sound path, and cross-talk attenuators or re-design of the ductwork may be found necessary. A similar situation occurs when ductwork leads to an extremely noisy location, e.g. external noise or a dance hall. This noise may pass through the system to a critical area. One particular case where noise may be easily introduced from outside is where small units are fitted on external walls. If there is an opening to the outside to allow air in or out of e.g. a fan-coil or heat-pump unit, this may be a significant path for noise between outside and inside. This is apart from any noise which the unit itself may make. Attenuated airways must be carefully designed. If the airpath is too narrow, devious and lined with absorbent, the fan performance may be affected by an excessive pressure loading. Too large an opening with inadequate lining will not give enough attenuation. Therefore, a careful balance is needed to achieve noise control, without an adverse effect on the performance of the unit. Account must be taken of such possibilities when attenuation is selected. The effect of mechanical noise on neighbours is a further consideration. In practice, air-intake and -discharge louvres, flues, cooling towers and air-cooled condensers are common noise sources. Disturbance may also arise from other forms of servicing, such as car parks, loading bays and refuse disposal.

The Use of Masking Sound

There are cases (e.g. when voices must not be overheard in adjacent rooms) where some control is possible by introducing other more acceptable sound in the receiving room, to provide masking (see p. 272). We will point out in the next chapter that there are preferred forms of background sound. But, of course, there are limits to how far sound can be used to 'cover up' noise without the masking sound itself becoming intensive. Even so, the manipulation of background sound levels and spectra remains an important tool for noise control. This is particularly so where improvement in insulation or attenuation between source and receiver is impractical. This is discussed further on p. 200.

Vibration Isolation

The isolation of vibration will now be considered. First, we shall distinguish what we mean by 'vibration' *vis-à-vis* 'noise'. By vibration we mean the movement to and fro of a structure or any other solid body caused by some alternating force, such as an out-of-balance rotating piece of machinery. This alternating force may cause the source to vibrate (i.e. the machine to vibrate, if it is a machine) or it may cause the structure to which the machine is fixed to vibrate, and in turn it may cause some other part of the structure to vibrate because of forces transmitted through the structure. From this point of view the ground is considered part of the structure and thus a vibrating machine, an underground railway or pile driving outside a building may cause vibrations in the building because of transmission through the ground.

Vibration may have four effects. First, it may cause damage. Secondly, it may be annoying to people. Thirdly, it may interfere with work, e.g. precision instruments, and lastly, it may cause noise.

It can be said at once that damage due to vibration is rare, and long before vibration becomes damaging it will be intolerable to the inhabitants. For example, an amplitude of vibration (i.e. maximum displacement from the mean position) of a floor of 0·05 mm at a frequency of 10 Hz would be annoying to a person standing on the floor, but it would have to be at least twice that amplitude to cause minor damage, unless indirect damage, e.g. from induced settlement of the foundations, occurred. It should also be remembered that much minor building damage, such as cracking of plaster ceilings, always occurs in buildings due to the normal movements of the structure resulting from the thermal and moisture changes. These cracks may often not be noticed until the building is being looked at carefully to see if there is any vibration damage.

Precision instruments such as microscopes and enlarging equipment involving high magnification cannot be used even with low vibration levels. Surfaces vibrating at frequencies within the audio range will act as noise sources and much of the vibration isolation required in and around buildings is directed towards reducing the noise arising from structure-borne vibration. Techniques for relating vibration levels to the resultant noise output have improved in recent years. But currently criteria for vibration isolation tend to be based on site experience rather than detailed prediction (see p. 176).

As with noise, treatment for vibration is based on an assessment of source, path and receiver. If we are dealing with vibration from an existing underground railway, assuming that the receiving position is very critical and no means of tackling the source is practical, the most appropriate method may be to isolate the room in question (or in some cases, the whole building) from the ground. But if a small pump is the source, it is much simpler to deal with the problem at source. Again, we must check all paths of vibration transmission. We might place anti-vibration mounts under the pump and motor assembly, but would be left with vibration transmission via the pipe-work (and perhaps via electrical connections). One bridge across anti-vibration mounts is often enough to ruin the isolation. We must therefore provide flexible conduit and flexible pipe hangers, to keep the vibration out of the structure.

Increasingly, the structure itself is becoming an important factor in determining the form of isolation. Whereas, in the past, buildings were more massive in construction, the increasing use of pre- or post-tensioning and lighter, more resonant building materials has changed the relationship between vibration and the structure.

Perhaps the most common mistake made in the practice of vibration isolation is the use of any resilient material, e.g. cork or rubber, for any vibration problem. Many problems have occurred because of this approach. Although frequently adequate by chance, many real problems have occurred as a result—vibration can easily become worse. At the outset, it is important to stress that any material required to provide vibration isolation must be carefully selected.

We will now look more closely at the mechanism of isolation. Vibration forces from machinery may be simple (as an up-and-down motion or side-to-side). But often they are complex, resulting in twisting and rocking motions. It is therefore necessary to start off with a clear idea of the likely plane of vibration. For the moment, we will restrict attention to vibration in a vertical plane, since this perhaps more commonly requires attention. If a machine sits on a spring on a solid immovable structure and it applies a very low frequency vibration force to the spring (i.e. it tries to force it up and down slowly), the stiffness of the spring will allow the force to be transferred to the base of the spring and no isolation will occur. But

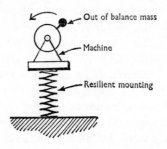

as the frequency increases, the inertia of the machine starts to produce forces out of step with the reaction of the spring (i.e. when the spring has just been compressed it is trying to push the machine up again, but the machine's inertia is trying to keep it going down). In this way the forces partly cancel each other and because the energy is stored, less force is required to move the sprung system. At one particular frequency, the opposing forces can cancel each other out At this point the system can be moved with very little force. Therefore, unless some constraint or damping is put on the system, the vibrational force will produce very large movements of the system. This is termed the resonant or natural frequency. If a vibration isolation system is in resonance the force transmitted is very large and isolation is reduced. As the frequency increases still further, the inertia of the machine restricts its ability to respond to the quickly changing directions of movement. As a result it becomes more stable and the movement and the forces transmitted to the structure are increasingly reduced. (The isolation also increases with softer springs, i.e. increased deflection under load.)

Therefore, it is only at approximately one and a half times the resonant frequency that any effective isolation is achieved (see Fig. 56). Although the transmission of forces decreases with higher frequency, it is important to remember that higher frequencies are associated with higher speeds of running and the actual forces to be isolated are likely to be higher. Therefore, isolation is not necessarily best at the highest frequencies.

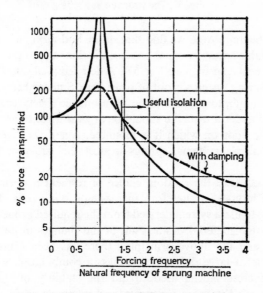

Fig. 56. Vibration transmission characteristics

From the above, for the case of a machine mounted on a massive non-compliant structure, vibration isolation can be achieved by careful reference to the resonant frequency of the mounted system. Fortunately, this is simply related to the deflection of the springs. Knowing the frequency of vibration of the source (e.g. from its running speed) we can then select a sprung system with a resonant frequency of no greater than one-third of the vibrating frequency, on the basis of deflection under load (stiffness) from the manufacturer's catalogue. We may need to incorporate some damping in the system to avoid problems during running up and running down through resonance when the machine is started and stopped.

Unfortunately, since modern buildings are less massive and more

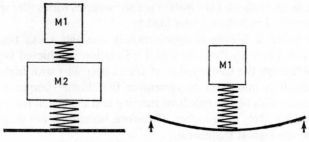

Fig. 57. The structure as a spring

prone to being moved, we find that, more and more, we are dealing with a double sprung system, because the floor under the machine is also acting as a spring (see Fig. 57). A detailed analysis of double or multiple spring systems is out of place here. Sufficient to say that, as a rule of thumb, to avoid unpleasant resonance effects, the deflection of springs under a machine should be at least three times the deflection of the floor on which it is mounted. Again, as a general rule, for all multiple systems, springs nearest the structure should be stiffest.

In this context, the random choice of resilient material for anti-vibration mounts is therefore increasingly dangerous. It is most important to use a system derived from the required properties of the anti-vibration material. As a result of the changes in building construction and the above principles, there has been a greater need for the use of large deflection springs in combination with inertia bases (concrete blocks used to aid the stability of the system). Although initially more expensive, such methods avoid very costly problems arising from inadequate isolation. For delicate instrumentation which is mounted on isolation in reverse, based on the same principles, it may be necessary to go to very specialist mounts (e.g. air mounts), which have a very low resonant frequency, to achieve high isolation.

Rather than attempt complex assessment of vibrating forces in other planes, in practice, it is often sufficient for most plant isolation to allow as much compliance in a horizontal plane as in the vertical plane (as with most springs and rubber-in-shear mounts). Of course, there will be special cases where more detailed analysis is required e.g. for the case of rocking in laundry equipment.

Isolation of buildings and rooms follows similar principles to those set out above. But a caution is necessary when the vibration source is not continuous. Whereas a continuous source will build up a steady state within the structure, the passing train or the driving of piles will be transient so that the behaviour of the structure is different—in particular, damping tends to be less effective. Again the principle guideline is to avoid resonance.

When buildings, plant, or apparatus are isolated, other practical problems are often introduced. For example, with buildings, wind loading will usually mean that lateral restraints must be provided. To avoid flanking, all service connections must be flexible. Scope for building settlement must also be allowed.

With other systems isolated from the structure, the effects of thermal movement, e.g. on pipework, can be troublesome. Pipe hangers and isolating sleeves must allow sufficient scope for such movement. Stability of spring systems can also be troublesome. Allowance must be made for flexible connections which are linked to rocking machinery and are therefore subjected to additional wear.

Well designed vibration isolation, therefore, involves not only achieving a reasonable degree of isolation and avoiding resonance, but also following the principles through down to the last detail and finding practical means of dealing with other problems arising as a result of isolation.

7

Criteria for Noise Control

Noise, defined as unwanted sound, affects people in a number of ways. It can be loud enough to cause temporary or permanent loss of hearing immediately. More often, hearing damage arises from exposure to high noise levels over long periods. Noise can interfere with warning signals and a wide range of communications, e.g. by speech. It can intrude, distract and annoy. We will see that buildings can also be too quiet.

It is essential, in modern building, to set clear targets for noise control. We see direct financial loss resulting from under- or over-design for noise control. In the noisiest areas, claims may be made in compensation for hearing damage. Noisy hotel bedrooms or offices are more difficult to let. Recording studios cannot function properly with too much noise. Expensive remedial procedures are often necessary to protect local residents from industrial noise or noise from leisure centres. There are numerous other effects of noise which are more difficult to assess directly, but can be loosely termed 'loss of amenity'.

Broadly, noise criteria can be grouped into two main classes. The first specifies the conditions required at the listener. The second specifies the reduction in noise (insulation or attenuation) required to achieve these conditions.

A wide variety of scales of measurement have been proposed to satisfy practical needs. Any confusion between these can be avoided by building up a clear understanding of the criteria which are in common use, and the context in which they are applicable.

NOISE AND PEOPLE

Individuals vary a great deal in their reactions to noise and vibration. Consequently, in parallel with other environmental criteria—heating, lighting, ventilation—there is a range of acceptable conditions, rather than a set of absolute standards. But the growth of experience in applying noise criteria has enabled practical targets to be established. Even so, the difference in reactions to noise is still there, and we meet extremely sensitive or insensitive people.

The main elements of noise which concern us are its level, frequency content (or spectrum shape), variation with time and any meaning or information implicit in the noise (e.g. intelligible speech or a dripping tap). Most criteria tend to be related to noise level and, usually, spectrum shape. A number cater for noise which varies with time. But the meaning or information in noise is extremely difficult to quantify and few criteria take account of this, although some have been proposed for particular sources.

For relatively steady background noise, the most common scales are the dBA scale and the Noise Rating Curves (e.g. NC, NR or PNC curves). For more variable noise, a statistical assessment is made, usually involving noise levels exceeded for a certain percentage of the time (e.g. L_{10}, L_{50}, L_{90}) or a measure of the equivalent steady sound level (L_{eq}). These are normally based on the dBA scale. Speech Interference Level (SIL) and Articulation Index (AI) are scales for specific use with speech. We may also meet scales developed for rating aircraft noise (usually PNdB and NNI). Each is briefly discussed as follows.

The dBA Scale

An A-weighting is a series of corrections (or when measuring noise, a weighting network on the sound level meter) which attenuates low frequencies and very high frequencies: e.g. noise at 63 Hz is weighted by 26 dB; at 8000 Hz the reduction is 3 dB. Between 1000 Hz and 5000 Hz, the noise level is slightly increased by the weighting (see Fig. 58). It has been found that if such a weighting is used when

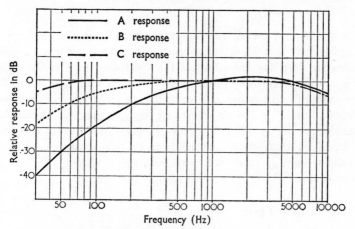

Fig. 58. A-, B- and C- weighting curves

noise is being measured, the resultant total level referred to as a level in dBA gives an answer which is a very good indication of the loudness for most common noises. Details of the weighting and also B- and C-weightings (which are rarely used for design criteria) are explained further in Chapter 9.

By applying an A-weighting network to the overall sound pressure level which is being measured by a meter, we achieve a single reading in dBA which gives an immediate assessment. The dBA scale finds favour partly because of convenience (we can refer to a single figure measured by relatively cheap instruments) and partly because it has proved to be as reliable a guide as many more complex criteria for a wide range of practical instances. The dBA scale has proved to be

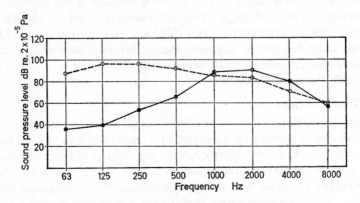

Fig. 59. Two noise spectra with the same dBA rating

the single most useful scale for rating noise over a wide range of applications. It is important to remember that a single figure in dBA does not give any indication of the spectrum shape of the noise, Fig. 59 shows two very different noise spectra which amount to the same overall dBA rating. But where the spectrum shape of particular types of noise is known, there are many situations where dBA ratings are highly practical. A lot of legislation on noise is written around the dBA scale. This tends to be related to variable noise sources (such as road traffic or construction sites) as will be discussed shortly.

Noise Limiting Curves

Closer specification of sound pressure limits at different frequencies is obtained using noise limiting curves. These take the form of a family of curves which again relate broadly to the equal loudness contours. Noise limiting curves are normally used as specified maximum sound pressure levels permissible in each octave band. The rating of any noise is then determined by the point on its spectrum which comes highest relative to the criterion curves.

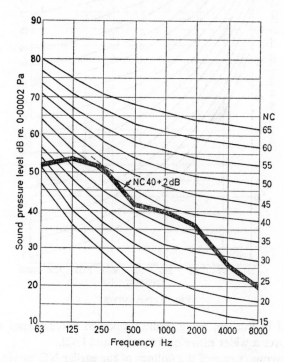

Fig. 60. NC curves, with an example of their use

There have been a number of such curves proposed. The Noise Criterion (NC) curves were originally developed in the USA, based on octave band centre frequencies which have since been superseded. Nevertheless, after adjustment to the current standard octave bands, they have been used extensively in specifications for noise control in buildings. They are shown, with an example of their use, in Fig. 60. To avoid doubt when interpolating between rating curves, we would not refer to e.g. NC42, but rather to NC40 + 2dB.

Noise Rating (NR) curves—see Fig. 61—were subsequently

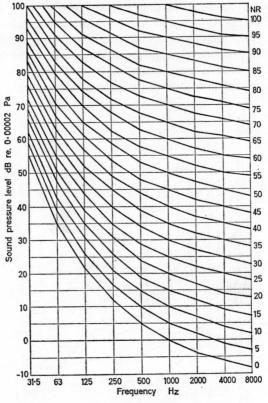

Fig. 61. NR curves

developed in Europe, using slightly modified curve shapes and extending over a wider range of frequency and level.

To overcome some of the failings of the earlier NC curves and to take some account of the fact that an NC curve when generated as such is not a pleasant sound, a further development—the PNC curves (see Fig. 62)—was made in the USA.

A number of other developments have been suggested, but none has become as well established as the above to date. It has been common practice to use noise limiting curves as maximum permissible limits and not as target spectrum shapes. As such they again allow a wide range of noise spectra to produce the same overall rating. There is an increasing tendency to allow for the preferred spectrum shape of the background sound rather than limiting loudness. A number of target spectra for broad-band sound have been

160

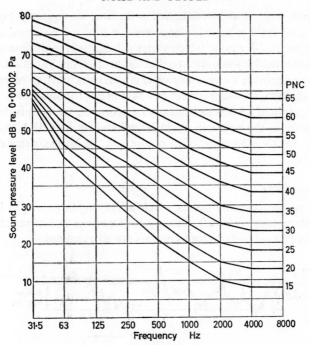

Fig. 62. PNC curves

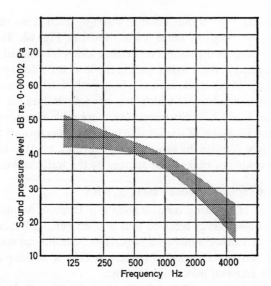

Fig. 63. Preferred sound spectrum (e.g. for speech masking)

F

161

proposed, most of which relate well to the zone we suggest in Fig. 63.

The dBA scale or noise limiting curves in the form discussed above cannot be used for noise which varies substantially with time unless complemented by some statistical assessment of this variation. Further, they provide no measure of characteristics such as repetition or recognition, although corrections have been proposed (see p. 173). For example, if a lift is to be near to a bedroom, the permissible lift noise level in the bedroom might be reduced by 5–10 dBA to compensate for the awareness of its nature and the intermittency of the lift noise.

L_{10}, L_{90} (dBA)

Many of the common noises which affect buildings vary a lot in level. Road traffic, aircraft, railways, construction sites, occupational noise in factories, offices and kitchens are all examples.

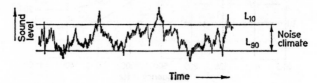

Fig. 64. Noise climate, L_{10} and L_{90}

Reaction to road traffic noise is strongly related to the 'noise climate' or the range of noise levels shown in Fig. 64. The upper limit of this range is the noise level in dBA which is exceeded for 10% of the time (L_{10}) and the lower limit is the level exceeded for 90% of the time (L_{90}). These levels are most frequently used as a measure of external noise conditions, which are strongly influenced by road traffic noise. L_{10}dBA is in common use as a simple representation of the potential intrusion of road-traffic noise and has formed the basis for much legislation on the subject (see p. 178). L_{90} tends to be used to apply to masking of other noise sources by the traffic noise.

Equivalent Continuous Sound Level (L_{eq})dBA

Another way of assessing variable noise is in terms of dosage, i.e. the amount of noise energy received over a given period. We can relate the energy in variable noise to an equivalent continuous sound level (L_{eq} in dBA) over the period of interest, usually a working day, and set limits in terms of permissible dosage.

This scale is well established for dealing with hearing damage and high noise levels generally, but can also be used for lower noise levels. Reaction to variable noise also depends on how much it fluctuates. The wider the fluctuation, the more annoying it tends to be. A refinement can therefore be made to L_{eq}, by adding a measure of the swing in noise levels. A scale of Noise Pollution Level (L_{NP}) has been developed on this basis.

Speech Interference Level (SIL)

Noise frequently interferes with speech communication e.g. in factories, on the telephone, in offices. The information in speech is largely contained in the consonants. These are weaker than vowels and occur in the higher frequency range. The interference caused by noise can be assessed broadly by checking the Speech Interference Level—the arithmetic mean of sound pressure levels in the octave bands centred on 500, 1000, and 2000 Hz. Maximum SIL can be specified in cases where speech communication is important.

Articulation Index (AI)

A more refined measure of speech intelligibility is the Articulation Index (AI). This measures the proportion of syllables intelligible: i.e. for 30% of syllables intelligible AI = 0·3. Clearly, the higher the noise level, the lower the AI value is likely to be. Fig. 65 shows a typical spectrum range for speech. The low frequency end of the

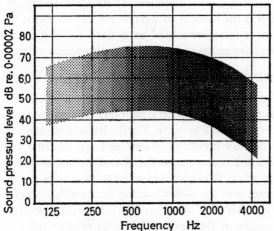

Note. Darker areas refer to those frequencies more important for the intelligibility of speech

Fig. 65. Typical speech spectrum zone

shaded zone represents much of the power, character and colour of the voice, but, as we have said, intelligibility is determined mainly at higher frequencies (in Fig. 65, the darker the area the more important it is to intelligibility). We cannot therefore simply plot noise levels over the speech zone and measure the proportion of area remaining above the noise. We have to attach more importance to the darker regions, and base AI on the proportion of the shading that is covered by the noise. Criteria for speech communication and privacy can then be developed on this basis (see p. 174).

Perceived Noise Level (PNdB)

This is a rating based on subjective response to 'noisiness' rather than loudness—there is a significant distinction between the two in that noisiness makes some allowance for information content. It has found use mainly as a measure of aircraft noise. The derivation need not concern us here. However, when met with PNdB figures for aircraft noise, it is useful to be aware of the approximate relationship with the dBA scale; i.e. level in PNdB $\simeq$ level in dBA $+ 14$.

Noise and Number Index

As a measure of the overall exposure to aircraft noise, the Noise and Number Index takes into account not only the level of noise but also the number of aircraft flying over.

$\qquad$ NNI = average peak noise level in PNdB $+ 15 \log_{10} N - 80$
$\qquad$ where N is the number of aircraft per day or night.

NNI contours have been used as a basis for definition of areas eligible to receive grants to improve sound insulation around airports.

SOUND INSULATION AND ATTENUATION

Our second type of criterion refers to degrees of noise reduction required to achieve the desired conditions. The scales normally used have been outlined in Chapter 6. For the sound insulation of building elements, such as partitions, floors, ceilings, external walling and roofs, doors and access panels the Sound Reduction Index (SRI) is most appropriate. It is common practice to specify the mean SRI, i.e. an average over the 16 third-octave bands with centre frequencies 100–3150 Hz. But it is often necessary to set out specific requirements in octave or third-octave bands to provide full specification for control of noise from specific sources. An alternative is Sound Pressure Level Difference, i.e. the actual difference in level between one room and the next without correction for partition area or room absorption. But use of this makes it more difficult to tie down component performance.

For specific sources it is often adequate to refer to sound insulation in terms of dBA.

Attenuation of noise, by screens, within ducted or piped services, at the source or at the receiver (in the case of hearing protection) is generally given as a dB reduction in each octave or third-octave band of interest. Impact noise is mostly rated by grading curves or received sound pressure levels resulting from the use of a standardised impact source on the floor above.

THE SETTING OF DESIGN CRITERIA

We can now consider the application of these scales and recommended standards to the activities which occur in and around buildings. We will start with the highest noise levels.

Damage to Hearing

Noise loud enough to cause immediate damage to hearing will not normally occur in buildings. But prolonged exposure in the noisy areas can result in damage, e.g. in factories, plant rooms or discothèques.

One suggested standard is a maximum exposure to continuous noise of 90 dBA over a working day of eight hours. If the duration of the exposure is reduced, a higher noise level is permissible—an additional 3 dB for each halving of the duration (see Fig. 66). Many of the noises likely to cause damage are not continuous. For example, machinery in a factory may be switched on and off several times

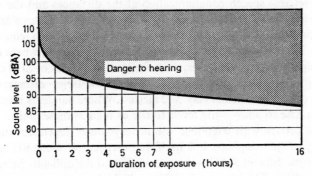

Fig. 66. A proposal for noise exposure limits to avoid hearing damage

165

during the working period. To rate this situation, L_{eq} can be used, again with a target maximum of 90 dBA (L_{eq}) for an eight-hour day. The value of L_{eq} can be built up by summing the dose during each period throughout the day. The chart in Fig. 67 can be used to do this. The reader must refer to current standards.

Of course, the specification should be referring to noise levels at the normal ear position of the receiver.

Hearing damage resulting from amplified music is not as well documented as damage from industrial or machinery noise. In the absence of comprehensive evidence, there has often been conflict between the noise levels favoured by the participants and those considered by many to present a hazard to hearing. Limits have been applied to peak noise levels (e.g. 110 dBA). But the indications are that these should be complemented by limits on the overall noise dose, based on the use of L_{eq} or a similar rating. This would need to be related back to the period of exposure. Clearly, the performers and staff of recording studios are more vulnerable than pop fans at individual concerts because they are exposed to the high sound levels for longer periods on a regular basis.

Speech Interference

The dangers of high noise levels are not all associated with the very loud noise that can cause hearing damage. At much lower levels, noise can interfere with warnings of danger, such as fire alarms, the sound of approaching traffic in a car park or urgent spoken warnings on building sites or in factories. At still lower levels, it can interfere with normal speech and telephone use.

Table II gives Speech Interference Levels (SIL) which will just allow reliable speech communication at the distances and the voice levels indicated. The figures in the Table apply when there are no reflecting surfaces nearby and the talker and listener are facing each other. This will not always be the case. In some rooms, where there are helpful reflecting surfaces, the noise levels (in terms of SIL) could be a little higher than shown, but only for the larger distances between talker and listener (say > 2 metres). Where danger is involved, it is unwise to allow noise levels to rise as high as these limits.

Interference with telephone use is very much dependent on the quality of the telephone link itself. In practice, telephone use begins to become difficult at SIL values of 45–50 and the difficulty increases until an SIL of 70–75 is reached, beyond which conversation becomes unsatisfactory.

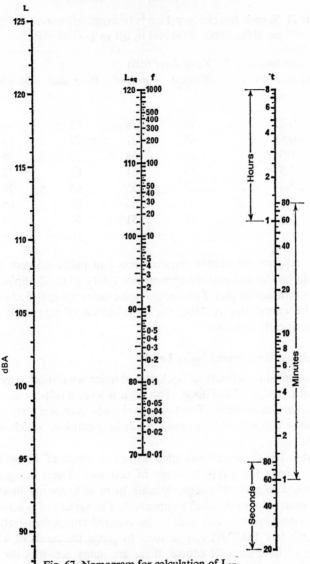

Fig. 67. Nomogram for calculation of L_eq

For each exposure connect sound level dBA with exposure
duration t and read fractional exposure f on centre scale.

Add together values of f received during one day to obtain total
value of f. Read L_eq opposite total f.

Example: 96 dBA for 1 hour   f = 0·5
         85 dBA for 7 hours   f ≃ 0·3
         Total f ≃ 0·8   ∴ L_eq ≃ 89 dBA

ERRATUM

Fig. 67, as printed, is inaccurate owing to the
left-hand scale (of dBA) being in the wrong
position: this scale should be 7mm to the left
so that the three scales are equally spaced.

Table II Speech Interference Levels (average of octaves centred on 1000, 2000, 4000 Hz) in dB re 2×10^{-5} Pa.

Distance between talker and listener (m)	Voice level (dB)			
	Normal	Raised	Very loud	Shouting
0·25	70	76	82	88
0·5	65	71	77	83
1·0	58	64	70	76
1·5	55	61	67	73
2·0	52	58	64	70
3·0	50	56	62	68
5·0	45	51	57	63

Speech communication requirements and public-address systems should be checked and the appropriate limits on noise applied. But, it is worth noting that if maximum noise limits are specified in terms of NR curves (see p. 160), the specification of maximum SIL is automatically included.

Maximum Background Noise Levels

Setting maximum limits on background noise is a critical stage in the acoustic design of buildings. Too high a level results in poor conditions and complaints. Too low a level leads to unnecessary expense on noise control, and can even result in conditions which are too quiet.

Table III gives recommended values in terms of Noise Rating (NR) curves for a typical range of activities. These are generally expressed as a target range, because in most cases maximum and minimum values are equally important. The variety of potential use of building space is too wide to be covered comprehensively here. Nevertheless Table III can be used to gauge the needs of a broad range of functions. Of course, there are many activities for which criteria may vary greatly. For example, kitchens in dwellings should not be so noisy as large industrial kitchens (perhaps NR35 rather than NR45). Criteria for music rooms or operating theatres differ widely. Similarly, sports and other leisure activities can often function well in a wide range of background noise conditions. It is therefore always important to examine the case in question and relate

Table III Recommended range of values for relatively steady background sound levels

		NR
Listening to music/speech*	Concert halls, large theatres, opera houses	15–20
	Multi-purpose halls, small theatres	20–25
	Courtrooms, conference rooms, debating chambers	25–30
	Small rooms, e.g. classrooms	25–35
Sleep and rest	Bedrooms, living rooms, lounges, hospital wards	25–30
Study and consultation	Libraries, executive offices, surgeries	25–35
Eating	Restaurants, dining rooms	25–35
	Cafeterias	below 45
Desk work with telephone use	Standard Offices	30–40
Repetitive tasks needing limited thought	General offices	35–45
	Large kitchens, noisy computer suites	below 50
Specific ancillary areas	Waiting	35–40
	Dressing/changing	below 40
	Circulation	30–40
	Toilets	35–45
	Car parks	below 55
Out of doors	In the street, in shopping precincts	below 55 (lower if external background levels are low)

*Studio recording and audiometry are normally subject to specific client requirements—NR20 or less.
Notes. Minimum levels refer in particular to middle and high frequencies. Levels must be adjusted downwards where the background noise contains repetition/rhythm, information, pure tones, etc. (see p. 173).

criteria to basic needs such as the need to communicate, the extent of concentration required or the need for privacy.

In specifying the limits on background noise, it is important to give the location of measurement, which will normally be at listening height (1 to 1·5 metres) and at least one metre from room boundaries. An NR rating does not cover the character of the noise, e.g. impulsive or repetitive. These aspects are difficult to specify without developing conditions which are too complex to be practical. It is more realistic to call for conditions free from undue tones, rhythm or similar distracting information content. Even so such conditions are difficult to interpret accurately.

Minimum Background Sound Levels

While most noise control is concerned with 'keeping things quiet', in some cases it can be too quiet. A quiet location is also one which is vulnerable to noise because, if a room is generally very quiet, noisy events such as a voice in an adjacent room, or an aircraft flying over, will be easily heard in it. But if the background sound level is raised and is relatively steady (as might happen when switching on mechanical ventilation), many of the noises which would have been heard in the quiet room are no longer perceptible.

In buildings in rural areas, the smallest noise will be heard. But the presence of the sound of e.g. a waterfall outside the window is often sufficient to stop such noise being heard. In many buildings the masking provided by background sound is very necessary to maintain the required standard of privacy. Provided the background sound is not of itself distracting, it is likely to be preferable to the noise which it masks. The quieter the room the more protection it needs and sound insulation can be very expensive. For example, in a recording studio, very quiet conditions are essential. Studios are therefore vulnerable. This cannot be avoided easily. But typical studio background sound levels in a hotel bedroom would invite noise problems from adjacent areas unnecessarily.

Therefore the minimum values in Table III are of considerable consequence. Since many privacy and distraction problems resulting from very quiet conditions occur at middle and high frequencies (> 400 Hz), the data for minimum levels refer primarily to this range.

Just as there are practical difficulties in reducing noise below a specified level, it may not always be easy to achieve reliable minimum sound levels. The practical aspects of background sound control will be discussed more fully in Chapter 10.

Road-Traffic Noise

Road-traffic noise getting into a building need not necessarily be subject to the same target as the noise from any mechanical services in the building. In critical cases (e.g. studios, auditoria), the road-traffic noise should not be perceptible as such and should be well below the steady background noise level. But in less sensitive areas (e.g. offices) and at low frequencies, where speech interference is less likely, the contribution of road-traffic noise may well be higher than that of mechanical services. Where both sources occur, the additive effect will, of course, push up the over-all level. But, at worst, this will result in noise levels 3 dB more than the higher of the two. Where this is considered particularly significant, allowance should be made in advance.

Table IV provides a guide to the selection of road-traffic noise levels relative to criteria for steady background noise (e.g. from services). If road-traffic noise varies a good deal and one can hear brakes, fast acceleration, car horns, etc., a further reduction of 5 dBA L_{10} may be needed. For tall buildings the difference in noise exposure between lower and higher floors may be small—this is because the higher the building, the greater is the area from which the noise is received. However, at low level, the noise is more erratic; at high level (i.e. over six storeys high, usually) the character of the noise is less intrusive. In some instances, an increase of up to 5 dBA is allowable at the top of a tall building, for that reason.

Aircraft Noise

Despite the very loud noise from aircraft, and the wide areas affected, it is difficult to relate the only commonly used scales (see p. 225) directly to building design. PNdB ratings are available for the majority of common aircraft and NNI figures have been used to assess the exposure around airports. But permissible aircraft noise in buildings is dependent on the particular aircraft activity, its variation and the use of the building.

As a rule of thumb, it appears to be desirable to limit peak aircraft noise levels to within 5–10 dBA above the background sound level, where sleeping is involved. 10–15 dBA is common and workable in many cases. Peaks above background up to 20 dBA are feasible for buildings such as offices, without undue disturbance. But a target maximum of 15 dBA above the background sound level is recommended. In auditoria and studios, aircraft should not be heard. This is a severe limitation calling for costly sound insulation. Siting of such building in an area of high exposure to aircraft noise is to be discouraged.

Railway Noise

Specific scales of railway noise assessment have not been fully developed. Desirable limits will depend to a great extent on the character of the noise as affected by the condition of the rolling stock, the track, the form of traction, brakes, horns, timetabling, etc. But, in the main, the limits on peak noise levels referred to above for aircraft seem also to apply to railway noise fairly well.

Table IV Suggested target levels for road traffic (L_{10}dBA) relative to specified targets for relatively steady background levels (NR), e.g. for mechanical ventilation systems.

NR	Road-traffic noise (L_{10}dBA)
20	20
25	26
30	33
35	40
40	45
45	50
50	55

Note. Some adjustments can be made to these values to account for the character of the noise (see p. 171).

Noise Leaving Buildings

Control of continuous noise leaving a building will normally be based on keeping below the L_{90} external background sound level. The presence of road traffic in metropolitan areas is helpful in relieving noise problems. Maximum limits to noise emission from relatively steady sources is usually given in terms of noise limiting curves (such as NR).

Music escaping from buildings is a much more difficult problem. Experience suggests that if the noise level is reduced below the steady external background, the rhythm and pattern of the music are often still perceptible and annoying to local residents. The same problem occurs to a lesser extent with industrial noise, noise from school playgrounds and many similar neighbourhood noises.

There is legislation limiting the emission of noise from buildings, as discussed on p. 178.

The Character of Noise

There have been a number of attempts to adjust criteria to allow for 'character' of noise. But these have necessarily been limited to very clear properties (such as 'impulsive' or 'containing tones') and cannot be expected to tackle the varying experiences and sensitivity of the listener. For example, some standards have suggested that tonal noise should be assessed 5 dBA higher than measured. Similarly, 5 dBA should be allowed for impulsive noise. Intermittency and duration may also be considered a basis for adjustments. But the breadth of subjective reaction to the character of noise is such that, if simple manageable standards are to be used, e.g. to deal with disputes over noise, there will always be cases which are not well served by them.

Sound Insulation

Sound insulation standards are often specified at the same time as the other criteria referred to in this chapter, because this provides a direct yardstick for the selection of building components.

The characteristics of the noise source should be considered carefully. The amount of background sound in the receiving room will also affect the degree of insulation needed. The detailed calculations used for obtaining the required insulation are covered in Chapter 9.

The use of the mean Sound Reduction Index is common and convenient. For most common noise sources, it is likely to be adequate. But there are still many cases where a more detailed frequency breakdown is required. If the source output is known, the specification should then include reference to the SRI over the critical range. Octave band SRI figures are usually adequate in such cases.

If mean values are taken, the range over which these apply must be specified. This is particularly important when alternative components are being compared. Some suppliers will quote the performance over the standard range (100–3150 Hz) as recommended. Others may use an alternative reference range. One such alternative is the 'speech range'—usually from about 400 Hz to about 2500 Hz. This normally gives a higher average. Clearly it would be wrong to compare components on such different data, although this may well happen if the range is not clearly stated.

We must bear in mind the type of construction which is capable of providing the required sound insulation, to ensure that it is compatible with the building design. Furthermore, the way in which the components are assembled must be considered. For example there

173

will be little chance of achieving more than 40 dB mean SRI for a partition if noise is able to pass easily through doors, lightweight ceilings, pipe apertures or similar flanking paths. At SRI figures greater than 40 dB, the flanking via the solid structure bounding the partition must be considered. These aspects should be covered by the specification.

It is important when setting criteria for sound insulation to include values for the building envelope, plant rooms and other structures needed to contain high noise levels as well as the normal specifications for demountable partitions. Some of the information on the noises may be difficult to obtain at that stage. If that is the case, an informed estimate of the requirements will be much better than omitting these items from the schedule.

Sound insulation of party walls and floors between dwellings is covered by building regulations. These will normally call for construction of party walls and associated structures to achieve a specified sound insulation performance. The reader is advised to refer to the regulations current at the time of reading (see also p. 178).

Sound insulation to control normal occupational noise is less variable. The majority of cases involve speech as the noise source (e.g. offices, interview and conference rooms, waiting rooms, courts, consulting rooms, toilets). Whilst other types of noise transfer must be calculated on their merits, the sound insulation requirements to control speech can be considered broadly in terms of privacy and lack of distraction.

Privacy

Although many activities in and around buildings involve a need for clear communication of speech, there are areas where this is undesirable if privacy is to be maintained. In some cases, both may be required. For example, an office worker in an open-plan area may wish at a particular time to talk without being overheard by his neighbour, yet still be able to communicate clearly with him at other times. Careful thought must be given to requirements of communication and privacy, as the implications are far-reaching, in terms of personnel location, mutual sound insulation and control of background sound.

It is difficult to be accurate about speech perception when patterns of speech and hearing ability vary so much between people. But we can obtain a broad guide from the Articulation Index (AI). Fig. 68 shows the relationship between AI, communication and privacy. We

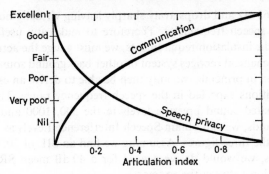

Fig. 68. Relation between Articulation Index,
communication and privacy

see that reasonable communication is possible even with low levels of syllable intelligibility. This results from the learned skill of filling in the missing pieces for ourselves. There are other factors, such as visual clues. Conversely, privacy is poor until intelligibility is low.

Although AI is a useful guide, it is difficult to check in practice. Consequently, we develop criteria for privacy from experience using sound insulation and masking sound levels. Insulation and masking are complementary in reducing the awareness of the content of speech. A very rough guide for relatively small rooms is as follows:

Privacy required

Mean SRI of Partition + SIL in receiving room

'confidential'	> 75
'good'	> 70
'fair'	> 65

Example

A row of individual manager's offices has a specified maximum background sound level of NR35. There is a requirement for 'good' privacy. Confidential privacy will be achieved by lowering voices on the few occasions when it is required. What standard of partition insulation is required?

This is a typical situation. But unfortunately, we do not have sufficient information. The NR35 is a red herring, since this is only the maximum limit. We cannot be sure that the actual levels will be this high. Unless we have also specified a minimum target as suggested on p. 170, we do not know what the actual levels will be. We could find that the noise level in the receiving room touches the NR35

175

curve at 63 Hz and drops away sharply leaving virtually no masking over the speech frequencies. Therefore to make any useful assessment of the insulation requirements, we must gauge the actual levels. If the mechanical services system or other background sound sources are known in principle, we may then be able to derive an estimate of the conditions expected in the speech frequency range. Then, from the estimated sound pressure levels in the 500, 1000 and 2000 Hz octave bands, we can obtain Speech Interference Levels as the mean value of the three figures. Assuming we find an SIL of 30, from our guidelines, we would derive a need for a 40 dB mean SRI for the construction between the rooms.

The higher the background sound, the lower the required partition performance. This can often be important, where insulation is difficult to achieve in practice, although there are limits on the level of background sound which can be accepted.

Impact Noise

Established criteria for control of impact noise are few. Most tend to relate to domestic building. Building regulations normally include maximum permissible sound pressure levels in dwellings, resulting from the use of a standard impact source mounted on the party floor above. The reader is referred to the current test methods.

For other building types, requirements will vary, depending on how critical the receiving area is. All areas with steady background sound levels below NR40 should be checked for potential impact noise from above. Simple floor surface treatments will normally be effective to control noise of footfalls affecting areas with NR25 and above. More critical areas are likely to require more elaborate protection involving special constructions to reduce impact noise.

Vibration

Detailed selection of vibration criteria applying to specific cases should involve a specialist. However, a guide to the effects of vibration on buildings and people is given in Fig. 69. It is useful to note that long before vibration reaches levels likely to cause damage to buildings, it will be intolerable to the inhabitants, although in some cases where buildings are set on weak ground some damage due to induced settlement is possible.

As yet, insufficient agreement has been reached on suitable criteria for vibration of buildings and their components by transient external sources, e.g. railways, pile driving. To achieve reasonable comfort

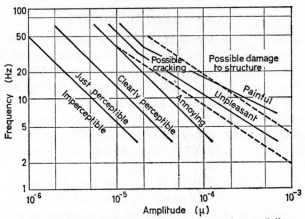

Note: This chart is directly applicable only for pure-tone excitation

Fig. 69. The effect of vibration on human comfort and building structures

standards, it is always possible to play safe by aiming to reduce vibration to levels which are hardly perceptible or imperceptible. But this can be very expensive, particularly when it is known that perceptible vibration of relatively high levels is accepted in many buildings without concern. Even so, it is always worth bearing in mind that vibration can be very worrying to people, and the presence of a ripple on the top of a cup of tea can be enough to initiate concern, particularly in the case of private dwellings.

Much of the vibration control applied to mechanical services installations is based as much on successful practice as on theory. But changes in the structural characteristics of buildings (towards lighter, longer spans, using pre-stressing and post-tensioning, etc.) have had far-reaching consequences on the methods. It is now more than ever necessary to consider the response of the building, in particular in relation to any likely resonance. Criteria used by mechanical services engineers still tend to be related directly to properties and performances of anti-vibration devices (e.g. static deflection under load, resonant frequencies) rather than to maximum structural vibration levels. It is always good practice to ensure that a check is made on the proposals for anti-vibration treatment to ensure that acceptable comfort levels are not exceeded, in terms of vibration or re-radiated sound. Where delicate instruments are involved, acceptable vibration limits should be obtained from the supplier, if possible.

177

LEGAL REQUIREMENTS

Legislation on noise is growing and being updated rapidly and is now increasingly influenced by European standards, so much so that it is not realistic to attempt a comprehensive survey of legal requirements and expect it to be up to date at the time of reading. Nevertheless, to offer the reader some guidance on the aspects of building covered by the noise laws, we will outline those areas which are affected (in the UK), at the time of writing (1977), without reference to the detailed demands of the law, which must, of course, be consulted in practical cases.

Apart from the availability of Common Law for cases involving noise, a wide range of building activity is now subject to legal control. The development of Noise Abatement Zones (areas designated for particular limitation of noise) will affect the broad planning of whole areas and the location of buildings. Legislation has required local authorities to plan noise zoning in advance, and a record to be set up of noise levels in the specified areas, as a reference for limitation of noise from buildings in these areas. Therefore, in dealing with new or modified buildings or change of use, conditions of planning consent may well include restriction on noise produced by a building, limits on the permissible noise exposure within the building or specification of sound insulation standards.

Where music and dancing are involved, the granting of licences is likely to be affected by the noise produced in the area by the activity.

There have been a number of compensation schemes provided for building in noisy areas (e.g. around airports, near new or modified roads). In these cases, grants have been made available to provide sound insulation and attenuated ventilation units to protect property which is exposed to noise levels higher than a specified value.

Building regulations set standards for airborne sound insulation for party walls and airborne and impact sound insulation for party floors, in dwellings. These will not only affect the construction of these walls and floors, but also the associated structure which carries flanking transmission. Regulations have allowed constructions which are 'deemed to satisfy' without requiring performance to be tested. The construction of separating structures between dwellings and refuse chutes have also been covered by regulations.

Industrial buildings will not only be subject to controls on noise emission as described above, but are likely to meet controls to protect hearing where very noisy processes are involved. Construction

178

sites will also come into this category. There are many noisy processes on building sites (see p. 277), which can damage the hearing of operatives. Owing to the common problem of noise emission from building sites, local authorities have increasing control over the way in which building (or demolition) may be carried out and advance planning of site procedures, with noise in mind, has become necessary.

8

Noise Control by Design

The control of noise is as much a matter of awareness as one of action. Many problems arise because noise is considered only when it occurs, rather than in advance. A lot of the early work in this field was remedial. But experience has led to increasing effort in the building industry to anticipate and control noise by design. This involves almost any aspect of building, from the choice of site or structural design down to the smallest detail—a door seal, or a small pipe bracket carrying vibration. Building designers must not lose the awareness of noise amongst all other pressures.

We will look at building design to pick out the main points which need attention at each stage. These vary somewhat between different types of building. The needs of particular building types are therefore discussed later in this chapter.

THE DESIGN PROCESS

Planning

We begin with the siting of buildings in relation to noise sources, other buildings and the landscape. Some progress has been made e.g. in zoning noisy industry away from housing. But in general, consideration of noise has not been well reflected in planning. Many of the old siting problems remain, e.g. the local community hall, where a lively evening is still often disturbing to local residents, a motor-racing track near housing developments or recording studios near underground railways. Recognising that many other important factors affect the choice of site, the cost and consequences of poor noise zoning are still often underrated. Careful planning can avoid noise and vibration problems before they arise.

Site Planning and Preliminary Design

Having chosen a site we must know, first, the type, extent, duration and character of the noise affecting it. Very often major decisions on layout and construction can only be based on a full survey and on realistic predictions for the future. An underground railway might

180

mean that all or part of the building must be vibration isolated. High outside noise levels may mean that sensitive areas have to be surrounded by less critical rooms. Aircraft noise may be enough to change roof construction radically, e.g. over a lecture theatre.

Surveys should be carefully planned to be sure that they give the information needed. For an office block, a daytime survey is often sufficient. For an hotel, night time is equally important. Misleading information can arise from measurement on early-closing days or when aircraft flight paths are not typical.

Predictions for the future are often awkward e.g. road-traffic volumes for ten to twenty years ahead cannot be certain. But where major changes are planned, the noise effects can be allowed for. Perhaps, for example, space can be left for doubling some of the glazing when a major road is to be built later. Even when major changes are foreseen, it is a mistake to leave out a site survey. Very often, a visit to the site (with noise and vibration in mind) finds important points which would otherwise have been overlooked.

Buildings often bring noise with them. It is important to check whether any restriction on noise from the site forms part of the planning consent. A record of the background noise level should be made before building. This allows the effect of new noise to be judged (see p. 178). Where 24-hour or late-night noise occurs (e.g. mechanical services, late-night dancing) a night survey is often necessary to find the lowest background condition. Noise which is not noticed in the daytime may well become annoying at night when background sound levels drop (see Fig. 70).

We must review the transmission of noise/vibration between the different parts of the building. It is, perhaps, unfortunate that, in the

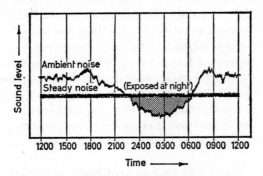

Fig. 70. The effect of a night-time drop in background noise level

181

normal way, sound is not represented on a plan. For example, the small rectangle labelled 'diesel generator set' looks so innocuous that there appears no very good reason why a bedroom window should not be located nearby. If, however, diagrammatic representation of this machine automatically carried with it a large black blot, say, 50 metres or more in diameter, signifying the noise it produces, there would be less chance of it being overlooked. It is impossible to stress sufficiently the value of taking one set of drawings, however preliminary, and marking up the noise-producing and noise-sensitive areas (see Fig. 71). Noise zoning must often be com-

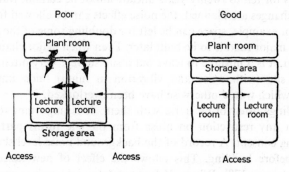

Fig. 71. Example of early planning of relation between
noisy and sensitive areas

promised with circulation, building form, costs and many other factors. But decisions about the location of rooms should be taken in the full knowledge of the implication on noise control.

To make sense of this procedure, we need a clear idea of acoustic design targets. These should be set as soon as the brief on accommodation is fairly well defined. Targets should not be limited to background noise levels or reverberation times but should extend to matters of privacy, vibration sensitivity, impact-noise control, etc. Setting comprehensive targets (as in Chapter 7) is useful in making one aware of any items so far overlooked during the site assessment.

When proposals for the building are taking shape, we must now look at where sound insulation or vibration isolation will be required —in particular, plant rooms, other loud noise sources (e.g. music rooms), the building 'envelope' and any floated or separated structures.

Very noisy rooms will tend to need heavy construction to help sound insulation. In some places (e.g. at high level) it may be better to use separate constructions to save on weight (which would have to be carried right down the building). Normally the structural engineer will need such information as early as possible. Mechanical plant is not usually selected in detail at this stage, estimates of sound power output have to be used to calculate the sound insulation needed.

The outside walls, roof (and sometimes floors) must be designed to give sound insulation sufficient to deal with noise from outside (or from inside travelling out). Small changes in construction can have large cost implications. It is expensive to overdesign, yet essential to obtain adequate insulation. Designs must therefore be based on the best possible information.

Where any separated structures are needed, special provisions (mountings, costs, effect on services, etc.,) must be settled at an early stage.

Preliminary designs should allow sufficient space for noise control. For example, the floor-to-floor height of a building has a strong influence on the space available above a suspended ceiling for services. If services layouts are cramped, noise is a likely result. The same applies to the size of plant rooms or builders' work shafts. Sound lock lobbies (see p. 270) are another example where space must be allowed in good time.

Scheme Design

As the design develops, the practical aspects of the features described above have to be dealt with. Specific selections for glazing, walling, roofing, and other major elements must be made. More attention is now given to items such as suspended ceilings and non-structural partitioning. For all of these the influence of non-acoustic aspects must be kept in mind.

For example, glazing specification will be strongly affected by e.g. cost, wind loading, cleaning, thermal insulation, natural ventilation, and perhaps smoke-venting requirements. A suspended ceiling may need to satisfy criteria of e.g. cost, stability, light reflectance, maintenance, control of spread of flame; and it will often need to integrate lighting and other services. All these influence the acoustic performance. Therefore noise control cannot be easily separated out at this stage. Alternatives must be checked against a wide range of criteria.

By now, mechanical services plant calculations can be based on detailed plant selection and ductwork/pipework layouts. System

183

attenuation, noise break-out and vibration isolation can now be tackled. Noise from lifts, document conveyors, computer, kitchen or office equipment, impact noise and speech and music sources should all be checked and proposals for speech reinforcement, public address, music or other background sound systems covered.

It is important to keep an eye on the logic of the acoustic design at this stage. For example, demountable partitioning systems are often weakened by noise passing through the suspended ceiling void above. It is an expensive waste of resources if either the ceiling assembly or the partition has significantly better insulation than the other. Doors are a common cause of similar weakness.

Detailed Design

Successful noise and vibration control depends very heavily on good detailing. A poor seal or a 'bridged' vibration isolator is enough to cause complete failure.

All sound-insulating construction must be detailed to be well sealed. This means careful selection of door or window seals, joint sealing strips or surface sealing (plastering or painting) or simply arranging components in such a way that these seals will be fully effective. Carefully chosen ironmongery often plays an important part in making a good seal. Equipment for vibration isolation, e.g. spring hangers, flexible connections or anti-vibration mounts, and volume-control devices i.e. valves and dampers, all need to be chosen carefully.

At this stage we need a full knowledge of the properties of the materials being used, and careful specification to ensure that those supplying components or doing the building work are clear as to what is required from them.

Supervision of Work

As with most aspects of building, it is not always easy to obtain results on site which match design intentions, even with careful specification. If those on site are not made aware of the importance of details at the appropriate time (i.e. when the job is being done) poor or misguided workmanship may well lead to weakness.

This is particularly true where the requirements are in any way very different from the normal run of events on site. A small gap around a pipe passing through a wall, very important as a path for noise, may not seem significant to the builder. Brickwork which is to be hidden from view above a lightweight suspended ceiling will not be carefully pointed (to achieve good insulation) unless this is expressly asked for and checked.

We also have to look out for the impractical method, which has been proposed, but for some reason does not work out and is replaced by an unsuitable alternative, without reference back to the designer.

To avoid these traps, supervision on site during key operations is invaluable.

Commissioning

It is interesting that a great deal of noise and vibration control is never finally checked (unless a problem arises). Without feedback of this type, there is a danger of continual overdesign or, at least, loss of useful experience for future work. Even at this late stage, conflicts may occur e.g. between air-flow quantities and noise. Many a scare over noise occurs when levels are checked before ventilation systems are balanced or before rooms are furnished. We must be certain that conditions are 'normal' before a final check is made. Specification of design targets should state the method to be used for checking at this stage and any tolerances (if any).

To summarise, noise and vibration can be controlled most effectively if the principles discussed in Chapter 6 are applied at the right time in the design process.

SPECIFIC BUILDINGS

Certain points tend to recur in particular types of buildings. A brief summary of major items affecting noise and vibration control is therefore given below for a number of common building types.

Auditoria

We will take auditoria to mean concert halls, opera houses, theatres, churches, cinemas, courtrooms, council chambers and lecture rooms.

First, all noise from outside and any source inside the building must be reduced to a very low level in the auditorium. Where noise contains any significant information, reduction must be even greater.

Any important auditorium on a town or city site should be protected from external noise by less critical areas planned around it. Where this is not possible, e.g. where a roof must be exposed to the exterior, a deep double construction may have to be used. Incoming noise, e.g. from a railway or aircraft may not occur often, or for long, but when it does, it may be during a quiet passage in a musical performance, a play or a film. In most cases, sound insulation will be designed to meet the criteria given in Chapter 7, Table III. But occasionally, e.g. in a lecture theatre exposed to aircraft noise, the cost of very high sound insulation may not be justified and insulation will be designed to ensure that the noise will not interfere with speech unduly (see p. 166).

Windows should be avoided if possible. When they are necessary (and they can be used safely only on fairly quiet sites) they should be as small as possible and well sealed, using single or double glazing depending on the noise climate.

Doors should be heavy and when they divide the auditorium from a noisy area, they must have adequate means of sealing. More often, the entrances need to take the form of a sound lock (see Fig. 107). Doors must not bang on closing.

If the site is close to sources of vibration such as a railway, on the surface or underground, structural separation of the auditorium may be needed. This is extremely costly and the need for it must be beyond doubt. Very careful design is required if this is the case (see p. 259).

Within the building, auditoria should be planned well away from rehearsal rooms, particularly where music is involved.

If these areas have to be adjacent and used concurrently at least 70 dB of insulation would be necessary and this could only be achieved by very special construction. Lift shafts should be kept clear of the auditorium.

All circulation spaces, foyers, bars, restaurants and any other rooms in the protective zone around an auditorium should have impact-noise-reducing floor coverings and absorbent linings.

Mechanical services noise control involves, in particular, the siting of noisy plant rooms and lifts away from the auditorium, providing adequate space for substantial attenuation, use of low air velocities and vibration isolation. In large auditoria there is often a conflict between the need to achieve a long throw for air whilst maintaining low air velocities. There is also a danger of external plant noise breaking back e.g. through smoke vents. Since external noise would also break in this way, proper detailing should deal with both problems.

Where auditoria are grouped together, mutual sound insulation requirements may well be extremely high. Noise in control rooms close to the seating must be carefully contained. In both of these cases, the relationship between insulation and background sound levels must be examined. In auditoria to be used for recording purposes, design noise levels will be relatively low and sound insulation problems are likely to be worse in these cases.

In courtrooms, noise levels should be sufficiently low to allow reasonable speech communication but not so low that quiet conference with counsel is heard clearly by the opposition. (See Chapter 7, Table III.) Where courtrooms are adjacent to one another, low sound levels require dividing construction to provide high sound insulation. Confidential privacy is needed between e.g. the courtroom and witness waiting or jury rooms.

The occupants of auditoria can themselves produce considerable noise. Careful furniture design can help but a lot of the noise, e.g. coughing in the theatre, or paper shuffling in council chambers, is impossible to control by design. Noise levels for mechanical equipment such as stage machinery, lighting gantries or projectors must be restrained.

Community Halls, Pubs and Clubs

Particular problems often arise in buildings where music and dancing is introduced. Often sited near houses, these activities can cause practical difficulties for noise control which should not be underestimated.

Apart from the level of noise produced, we also have to consider the information contained in the noise. The slightest suggestion of rhythm is enough to make the quietest sounds disturbing. These buildings are often sited in areas where background noise levels drop in the evening. Then the faint beat of a drum from a nearby public house is enough to cause severe disturbance. This is pointed out not to discourage the use of sound-insulating construction, but more to underline the scale of the problem. As a guide to the extent of control required, the target for noise reduction should be at least 5 dB (preferable 10–15 dB) below the background noise levels at the listening position. This should be so not only in terms of dBA, but, in particular, the low frequencies must be carefully checked.

Ventilation and fire-escape requirements in function rooms tend to lead to noise leakage. Large glazed areas are not helpful either. Open windows or roof lights are obviously not appropriate. Ventilation systems should not take the form of small window-mounted or roof-mounted extract fans, drawing out air which is replaced by natural leakage. Properly attenuated systems are necessary.

Fire-escape doors are often poorly sealed and held to the frame loosely by pushbars and espagnolette bolts.

Even if seals are provided and ironmongery arranged to maintain a good seal (whilst still allowing a quick exit) sound insulation is rarely good enough. Specialist acoustic doors with high performance are seldom compatible with fire escape. Their ironmongery and the sheer weight of them rules them out in most instances. The use of a sound lock lobby is often the best practical solution.

In some cases, the insulation requirements of the building enclosure are too high to be practical at reasonable cost. In such cases, it may become necessary to limit the level of noise produced in the function room. Automatic volume-control limits on amplifiers have been used in an attempt to achieve this—a measure which is not popular with participants.

Unfortunately, these gatherings often result in a noisy departure from the premises, e.g. car doors slamming, revving of motor bikes, a loud farewell or raucous singing. This sort of disturbance is unfortunately beyond immediate control by the building design, but remains an important source of annoyance to neighbours. As such it can affect the granting or otherwise of music and dancing licences.

Hospitals

Hospitals are heavily serviced by mechanical equipment. Apart from central plant (boilers, fans, cooling towers, power generators, etc.), there is a wide range of special equipment e.g. sterilisers and autoclaves. Noise in kitchens, e.g. from dishwashers, automatic mixers and refuse-disposal units and the normal impacts on crockery, cutlery and metal surfaces often spreads to sensitive areas. Lifts, trolleys and cleansing equipment, children's wards, laundries and loading bays are all likely to be noisy.

It is therefore very difficult to separate sensitive areas from noise without careful planning and detailing. Hospitals are a very important application for the principles of noise zoning (see p. 182). Whether the main noise sources are grouped together or spread throughout the hospital, their position *vis-à-vis* sensitive areas must be checked. It is bad practice, for example, to put a noisy mechanical plant at the bottom of a glazed internal court which is surrounded by wards. If all the major plant is grouped in a central station, the location must be carefully chosen. If it is not well separated from sensitive areas, plenty of space must be allowed for good sound-insulating construction and noise-control equipment.

The use of hard surfaces to allow easy maintenance and hygiene tends to lead to very reverberant surroundings. As a result noise levels build up, e.g. ventilation services in operating theatres, or the noise spreads more easily, e.g. along corridors. Hard floor surfaces are likely to lead to impact-noise transmission and surface noise.

Porous absorbent treatments are generally not acceptable in operating theatres or treatment rooms. Perforated metal pans with backing of porous material contained in a thin airtight bag have been used in many areas for general control of reverberant sound, e.g. in corridors. Occasionally carpet has been used in non-critical areas. Curtains or any linings which can be removed and washed are usually acceptable in wards and less critical areas.

Unfortunately, the choice of medical equipment is not often influenced by levels of noise produced during its operation. It is usually a matter of siting the equipment carefully and allowing for well sealed partitioning and doors. Careful siting is again the best defence against noise from service areas, kitchen, laundries and children's wards.

Noise from movement of e.g. equipment, food or people can be tackled by using easily maintained resilient materials, e.g. smooth flooring with resilient backing or underlay, rubber pads and buffers, plastic containers or door seals which limit impact noise on closing.

It is worth pointing out that much of this noise is easily heard if the sensitive areas are too quiet to start with. Continuous background masking sound is useful not only to counter noise, but also to improve privacy. Privacy is necessary in wards, consulting rooms and offices, in particular. These areas should have specified minimum sound levels. (See Chapter 7, Table III.)

Protection from noise becomes more critical in rooms used for testing hearing. Although much of this work is done using headphones, it is often necessary to carry out tests without them. An audiometric room may need to be constructed as a separated structure, if it is close to noisy areas. Otherwise, conventional heavy construction, e.g. plastered brickwork or blockwork, may be adequate. Noise control for the ventilation system, doors and windows often involves specialist designs.

In hospitals, two particular aspects of vibration need special attention. First, where any delicate instruments are in use, e.g. in laboratories, operating suites or treatment rooms, extra care may be needed to limit vibration. Use of specialist mounts on delicate equipment is frequently the most effective solution. Second, because patients are mostly in beds, more than usual consideration must be given to the horizontal components of vibration.

Consulting Rooms

See p. 203.

Hotels

Generally, the most critical areas in hotels are the bedrooms. We need to keep a careful balance between external noise, noise from adjacent rooms and services noise (see Fig. 72). Noise must not be of such a level or character as to disturb sleep. But too low a background level will allow disturbing noise to be heard more easily.

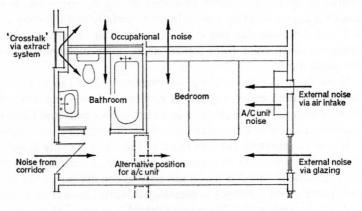

Fig. 72. Noise affecting a typical hotel bedroom

Windows, partitions and ventilation units should be chosen with this in mind. Often outside noise and noise from the next room is disturbing in character and must therefore be kept well down. The background sound must then be provided by the ventilation units. The level and character of ventilation noise therefore becomes extremely important. Plumbing noise is often disturbing in hotels —layout and fixing of piped systems should take this into account. Impact noise and 'cross-talk' between bathrooms and toilets (e.g. via a common extract-vent system) must be checked. Corridors should be treated with absorbent and impact-reducing floor coverings. Laundries and service rooms must be sited carefully.

Function rooms in hotels may well operate as multi-use areas,

190

needing good protection from outside noise and room acoustics appropriate for a wide range of use. Subdivision by folding partitions may involve sound insulation requirements. These areas produce noise when used for e.g. music or dancing. Car parking must be treated with consideration of the disturbance which can be caused when a large number of people leave a function. When conference rooms are included, speech privacy is essential.

Dining areas need some absorptive treatment and protection from kitchen noise. There is a conflict between the latter and ease of access between the two areas for serving, which can be overcome by careful use of doors and screens. In kitchens some absorption is useful to keep noise levels down. Allowance must be made for moisture and hygiene requirements. Kitchens and associated service areas should be kept well away from bedrooms.

Mechanical plant areas must be carefully sited and constructed.

Particular attention should be paid to siting of air-intake and -discharge louvres, vibration isolation of pumps, compressors, fans, etc., boiler flues and external cooling plant. Lifts and document conveyors should be checked for noise output.

Hotels are tending to include an increasing variety of facilities. A check through the schedule of accommodation should be made to pick out any other areas requiring particular attention, e.g. swimming pools, cinemas.

Housing

People in their homes are naturally very sensitive to disturbance by noise. It usually comes from one or more of three sources—external sources, the neighbours or central services, such as lifts and refuse chutes.

External noise problems are widespread and various. Road, rail and air traffic are the most common sources of trouble. Other examples are factories, playgrounds, dance halls, construction and roadworks. Legislation has increased controls on noise out of doors, mainly via local authority officers, who are now very much involved in noise zoning and planning.

A great deal can be done by selection of sites and site planning. Fig. 73 gives two examples of noise control by planning. Where separation from the source, by distance, external screening or shielding is not feasible, we have to rely on the sound insulation of the shell of the building. Therefore, in these cases, it is wise to avoid very lightweight constructions.

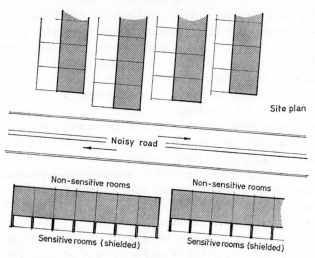

Site plan

Noisy road

Non-sensitive rooms   Non-sensitive rooms

Sensitive rooms (shielded)   Sensitive rooms (shielded)

Fig. 73. Layouts for housing to reduce exposure to noise

Under the Government Aircraft Noise Grant Schemes and the Noise Insulation Regulations, some householders (living very close to some major airports or near new or modified busy roads) have been given financial help to improve insulation. In these and other similar cases deep double glazing is usually appropriate and external doors need to be solid and well sealed. But then natural or mechanical ventilation must be provided without re-admitting or creating too much noise. Specialist units have been developed to allow this. With aircraft noise, we must give more attention to roof construction and blocking of unused chimney flues.

Noise from neighbours is very much affected by house plans and by party wall or floor construction. It is usual for building regulations to lay down minimum standards in the form of sound insulation performance for party walls and floors, and, in the case of party floors, impact noise isolation. Of course, these cannot be dealt with without reference to associated construction carrying flanking transmission, e.g. the inner skin of a common external wall, in the case of a party wall. By their nature, regulations are limited and can only be based on the needs of the majority. As a result, they cannot be expected to deal with all noise problems, where background sound levels vary widely (e.g. from the city to the country), subjective reaction is unpredictable and a wide range of noise sources are concerned. As a result, noise from neighbours, e.g. hi-fi, parties,

192

television, children and musical instruments is frequently heard, particularly where the source is close to (or even fixed on to) the party wall.

Traditional constructions of well pointed plastered brick (solid or cavity) or dense concrete normally satisfy the regulations, and increasing use is now made of lighter blockwork construction with a wider air space, dense concrete blocks or even separated framed and boarded constructions. In all cases, the division must be very well sealed by plaster and/or joint sealants.

Improvement in insulation between existing homes is rarely practicable or economical. Unless obvious holes or other weaknesses are clearly evident, little improvement is likely as a result of upgrading the party wall alone. This is because of the remaining flanking paths, which are very difficult to deal with in practice. Even where the direct path is weaker than the flanking path, considerable space and/or weight is needed to achieve minor improvements.

Lifts, refuse chutes, laundries and other central services are common sources of complaint. They should be separated from bedrooms and living areas, and detailed to avoid vibration and impact noise transmission into dwellings. Building regulations call for a minimum mass per unit area for walls between refuse chutes and 'habitable' rooms. Plumbing is noisy by tradition, although improvement is possible by careful sizing, layout and fixing of pipes and selection of quieter valves.

It is worth mentioning noise within houses. 35 dB nominal insulation is recommended as a minimum for internal partitions.

Industrial Buildings

It is useful to think of two categories of industrial noise—noise within the premises and noise affecting neighbours.

Priority must be given to reducing noise which could result in permanent hearing damage, and there are legal requirements to be met (see p. 178). This normally applies to people working very close to powerful noise sources. Reduction of noise at source, screening, enclosure (or, as a last resort, hearing protection) must be allowed for. These high noise levels tend to occur in the region where the direct sound is strong compared with the reverberant sound. But sometimes, dangerous noise levels may occur over large areas, partly as a result of a build-up by reflections from hard surfaces. Absorption by linings or suspended units may then become useful.

Below noise levels which are dangerous to hearing, we may still have enough noise to interfere with warning signals (a voice or a buzzer) and certainly normal communication by direct speech or public address. Incidentally, this is one of the dangers of ear defenders/ear plugs.

There are many noisy industrial processes, too many to list here. Sufficient to say, there are clear implications for the buildings which house them.

The layout of the process should take account of noise, and almost without exception some sound-absorbing lining or contents are helpful, even for speculative building where the use is unknown. Particularly awkward are the limitations on freedom of choice of materials, imposed by e.g. fire precautions, presence of oil mist (e.g. heavy industry), health requirements (e.g. packaging/bottling), chemical make-up (e.g. it may conflict with the product), impact damage and so on.

Protection of neighbours means keeping noise in. We need to provide adequate sound insulation for the building shell. This is not helped by many lightweight building methods. Weakness often shows up at large access doors (e.g. open roller-shutter doors), ventilation or smoke-extract openings, loading bays and garages. Mechanical systems, e.g. dust-extract equipment, power generators and their flues, and external cranage are frequent offenders. Mainly outdoor plants, such as oil refineries and quarries, need careful siting.

In these cases, we must know the background sound conditions at the sensitive positions (often local housing) in the absence of plant noise. The effect of bringing noise to the area can then be checked (see p. 181). It may be that the local authority has set down noise limits at the boundary of the site.

In summary, industrial building must take account of planning of the site, process layout, sound insulation and absorption, attenuation at openings and vents, and, perhaps, screening. Where offices are built into the factory, internal sound insulation also becomes relevant.

Libraries

Distraction by noise often disturbs use of libraries for reference or study. Siting and control of incoming noise must be considered in the usual way. Precautions should include: floor coverings to reduce surface noise and, in rooms above, to reduce impact-noise transmission; absorbent room linings to keep noise levels down; resilient buffers on chair and table legs and zoning of audio or copying equipment. Again minimum background sound targets are appropriate, to avoid conditions becoming too quiet (see Chapter 7, Table III).

Music Rooms

Apart from the level of sound produced in music rooms, the rhythm and tone of music is an added annoyance to the unwilling listener. Therefore containing noise becomes most difficult. High insulation values are needed for the construction separating the source from the listener.

Usually, we do not want the musicians to be disturbed by incoming noise. The sound insulation is therefore of benefit in two ways. But when music rooms are lined up in rows e.g. as practice 'cells', insulation standards must be extremely high. Perhaps in ordinary schools it is less realistic to aim for very high insulation values (50 dB nominal is often acceptable). But, for example, in a college of music or rehearsal suites in concert hall complexes, it may well be necessary to build separate floated constructions to achieve 65–70 dB nominal.

Careful planning, e.g. separation of music rooms by service ducts, instrument stores and the like, is one way of easing the problem. But detailing of doors, windows, ventilation and impact-reducing floor coverings remain essential.

To keep the sound of music in, it is a useful discipline to add 10–15 dB to the insulation requirements which would be calculated on the basis of sound pressure level difference above. This takes account of the information in the sound and avoids underestimating the seriousness of the problems.

Offices

Where people are working fairly close together, as in offices, we are mainly concerned to avoid distraction, loss of communication or simple annoyance due to noise. Adequate privacy must also be provided.

Individual Offices

In cellular offices, provided we have dealt with external noise, sound insulation of partitions and background sound are the main areas of concern. These two must be carefully balanced. The guidelines on privacy (referred to in Chapter 7, p. 175) should be applied here (see Fig. 74).

A number of particular problems tend to recur with cellular offices. For example, if demountable partitions are used below a

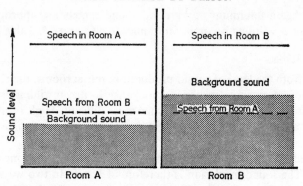

Fig. 74. Effect of background sound on privacy

lightweight suspended ceiling, noise transmission through the ceiling void weakens the insulation. The use of vertical barriers above the ceiling would need to be widespread to cope with the flexibility so often needed—partition positions are not often known in advance. Inserting a barrier when the partition is moved is very messy in practice and rarely well done. Apart from this, vertical barriers may well interfere with the ventilation system.

Use of a heavy sealed overlay on top of the ceiling is little better. With this method, access into the ceiling is difficult, air movement may be restricted, and suspension rods or straps often make installation difficult.

Usually the most practical solution, at the design stage, at least, is to make the ceiling heavier. There are problems where ceilings are perforated for sound absorption, by air-handling light fittings or simple return-air slots. But with careful selection of ceiling type and proper detailing, this method is effective. It allows flexibility of partition layout and may be useful as insulation against noise breaking out of ducts.

When pipes, ducts and conduit pass through partitions, the openings must be well sealed. Often this is difficult in practice, where pipes are close together. Access to make the seal is limited, and results are usually poor. With forethought, this situation is easily avoided.

Glazing reduces the overall insulation of many partitions. But doors are often much worse, particularly when poorly sealed. Sound leaks easily through open thresholds and via other air paths round the door. Large improvement (sometimes 10 dB or more) can be achieved with careful detailing.

196

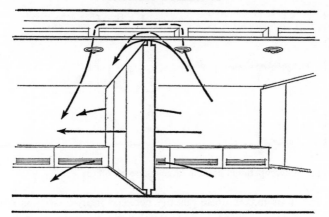

Fig. 75. Flanking transmission affecting partitions

Control of flanking transmission (see Fig. 75) can be expensive. It is therefore all the more important that full advantage is taken of background sound to complement insulation. Very often, considerable savings in the cost of achieving high insulation values can be made in this way. Where sources of background sound are not readily available, other means of introducing masking sound must be found (see pp. 150 and 272).

Shared Offices

If the work does not need concentrated thought or privacy, shared offices can be successful. But people together in small offices sometimes disturb one another. Distances for direct sound are short and there is little protection available besides screening. It is extremely unlikely that screening and absorption will be enough. Maintaining fairly high background sound levels can be helpful, but levels required for adequate masking are often too high to live with. This arrangement is therefore ill-advised, particularly if noisy office machinery is present.

Open-Plan Offices

It is difficult to be certain at what size an office can be termed open-plan. But we can consider open planning as an extension of the shared office, often reaching sizes catering for several hundred people in one area.

In general, larger offices tend to be more successful. This is because the activity of a lot of people creates steady background sound (see

197

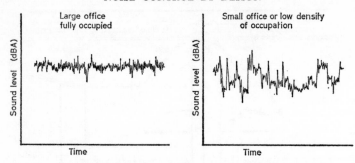

Fig. 76. Occupational noise in large and small offices

Fig. 76). This makes noise passing between individuals less notice-able. In smaller offices, occupational noise is more sporadic and therefore tends to be more intrusive.

It must be clear from the start that it is difficult to replace the mutual protection which full-height partitions give. We have to make use of every possible means to reach what are normally no better than adequate conditions. As a result, acoustics has been a major weakness in many open-plan offices. But with comprehensive design this need not be so (see Fig. 77).

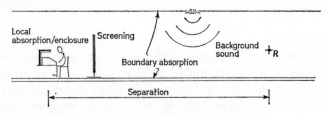

Fig. 77. Main features of noise control in open-plan areas

Requirements vary depending on the intended use of the offices. Naturally, there are fewer problems where the work done needs little privacy and mental concentration. We will deal with the cases where privacy is needed and distraction must be kept to a minimum. Sometimes group privacy is as important as individual privacy, i.e. people may be used to the noise within their own group, but are disturbed by adjacent groups.

We start by controlling noise at source whenever possible. This is difficult with people, but a lot can be done with office machinery. Telephone bells can be muted or replaced by more suitable signals. Typewriters can be properly mounted, even partially enclosed. Local absorption and partial enclosure are useful for telex, teletype and a number of noisy machines, provided access is not restricted too much. Certain machines (or people) are so noisy that they should be kept out of the open area altogether.

Zoning of noise sources away from sensitive positions is very important in open offices.

Screens are useful in the right place. The principles to follow have been stated in Chapter 6, p. 134. It is useful to consider how screens can be linked together or butted up to walls without large air gaps. Combining screens with desks is helpful, because they are then close to the source or receiver. But rows of part-height booths with reflecting surfaces, e.g. metal and glass, can give less protection than nothing at all. Sound builds up locally and occupants raise their voices as a result.

Reflected sound can be tackled by absorption (and to some extent diffusion). For the most part, sound is travelling in multiple reflections between the floor and ceiling. In general, ceilings should be absorbing. With modelling they can be less so. But reflecting ceilings tend to build up the overall level of activity noise to an uncomfortable level. Three-dimensional ceilings (e.g. egg-crate or V-sections) allow a lot of absorption to be introduced. But flat ceilings can be equally good provided the material is carefully chosen. With flat ceilings, at least 70% of the ceiling area should be absorbing and large reflecting areas, e.g. large light fittings, should be avoided. To reduce long reflections, absorption should be good at shallow angles of incidence.

Floors are usually carpeted. Selection for good absorption is important. A medium pile should be considered the minimum quality acceptable. Walls become relevant around the edge of the office area. Absorbent lining is useful, but awkward where glazed walls are concerned. If curtains are not acceptable, the use of blinds or some irregularity in the profile of the walls helps to break up bypass reflections around screens.

In offices containing few people or having a low density of occupation, background sound levels can be too low. As a result noise is heard more easily (see Fig. 74). This has been the major weakness in many open (and cellular) offices. In particular, we need sound over the middle and high frequency range to mask speech. Apart from the sound of normal activity, there are usually two alternatives—external noise and ventilation noise. External noise is often variable in level and is rarely of the correct spectrum after being filtered by the building construction. The sound from ventilation systems can be useful, particularly air noise at diffusers or induction units. But there is often a conflict between the required air flow and the sound level. Air volumes usually take priority. Variable-volume systems cannot be a reliable source of sound because of the variation in air noise. Sometimes, therefore, there is no suitable source of masking sound present, and other ways of producing the sound must be considered (see pp. 150 and 272). But background sound control however it is achieved should not be considered the sole answer to the problems of the open-plan office. Nevertheless, it can make a useful contribution in conjunction with the other procedures noted above.

The way that open offices are used after their design is often central to their success or otherwise. It may be helpful to provide the office manager with a few notes on how to organise the office to keep noise problems to a minimum.

Schools

A great deal of school building has made use of lightweight construction. As a result, a lack of sound insulation has been found to be a major limitation. Teaching, whether formal or informal, clearly requires good communication. Yet often it is disturbed by external noise, e.g. aircraft, road traffic, railways or by noise from adjacent teaching areas, music rooms, gymnasia, swimming pools, playgrounds or workshops.

If schools have to be sited in very noisy conditions, e.g. near airport runways, they should be built of heavy construction. With lightweight system building, if obvious weaknesses such as open windows and lightweight roof lights can be avoided, reasonable conditions can sometimes be achieved. But on the whole, school construction is currently too lightweight to achieve suitable protection on such sites. Sealing windows is often a useful improvement, but this creates a ventilation problem. Few schools can be allowed mechanical ventilation systems. Where they are used, 'cross-talk' through trunking, from the outside or between rooms must be controlled. Double glazing is only appropriate where the remainder of the external construction of the building is capable of matching its sound insulation. Roof construction is often weak in this respect and can be of particular concern with aircraft flying overhead.

The wide range of activities which occur in schools makes for a conflict between noise-producing and noise-sensitive activities. To allow flexible use of space, perhaps the most effective way to tackle

200

noise is by careful timetabling. But school buildings should be designed to allow a reasonable degree of freedom. Even in the face of severe cost limits, certain basic standards must be met if the school is to work at all. As a minimum, it must be possible to talk at a reasonable level in one classroom without disturbing the adjacent area or being overheard outside the room. The balance between partition insulation and background sound levels should be used to advantage. Detailing to avoid flanking around partitions is essential, but often forgotten.

Many areas will vary in noise production and sensitivity depending on their use at the time. Assembly halls may be used for sports training (usually noisy) or for quiet study (requiring protection). Music rooms must both keep noise in and keep it out at the same time. Activities such as 'crafts' are often neither noisy nor noise-sensitive. Where any pattern of use can be predicted beforehand, noise zoning is advised and sound insulation should be built into noisy or noise-sensitive areas as far as practical. Dining areas and circulation routes in particular should receive absorbent linings to keep down noise levels and restrict the spread of noise through the building. Impact-noise-reducing floor coverings are essential above ground floor level.

Open planning in schools presents new problems. It should be made clear that there are strict limits on the types of activity which can live together at peace in open-plan arrangements. Even then timetabling is again important and screening, background sound control, absorbent room finishes and careful layout have to be used to achieve just adequate conditions.

New forms of teaching involving audio/visual aids introduce more detailed design considerations in the form of carrels, booths and equipment noise control.

The effects of school playground noise on the school and on neighbours should be considered when the building is sited and the layout on the site determined.

Buildings for Sport

In most cases, sport is a source of noise rather than sensitive to it. From the roar of a football crowd to the clatter in a squash court, there is a wide variety of sporting activity which can disturb neighbours. Again, the information content in the noise makes it particularly disturbing. Outdoor sports, e.g. motor racing, need careful siting and screening from sensitive neighbours. But indoor facilities have increased in numbers rapidly. Very often these are grouped together with e.g. leisure facilities which are sensitive to noise. In these cases noise zoning and sound insulation become critical.

Often, the noise is a combination of calling and shouting by players or spectators and the noise of the sport itself, e.g. impact noise. Impact-noise transmission to adjacent rooms can be limited by structural separation. But much of the impact may show up as airborne noise radiated into the source room. In this situation, light-weight and resonant floor and wall constructions should be avoided and the use of sound-absorbent materials to limit the build-up of airborne noise becomes particularly helpful.

There is a risk of impact damage on soft absorbent surfaces, e.g. acoustic plaster or mineral-fibre boards on battens. Therefore, there are advantages in making use of perforated facings with absorbent backing. Absorbent materials should be chosen with more than usual care where humidity is high, e.g. in swimming pools and changing rooms. The chemicals used in pool water must also be considered, e.g. hypochlorous acid where chlorine is used.

Hard surfaces are usually required at low level in these areas for reasons of hygiene. Treatment is therefore often limited to the upper part of walls, balcony fronts and the ceiling. Omission of absorbent treatment during the design stages can result in expensive revisions, because treatment added after construction may require very expensive scaffolding.

Noise from indoor facilities may still break out, e.g. to nearby housing, if care is not taken to provide adequate sound insulation not only to the enclosure but also to services and exit doors, etc.

Studios

Very little noise can be tolerated in radio, television and recording studios generally. They range from the individual recording booth to the studio auditorium. But they have a common need for low noise levels, although levels may be higher when an audience is present.

Careful siting and shielding are essential. Broadcasting House, London is an interesting example of this (see Fig. 78). Low noise levels in studios mean that it becomes easier to hear incoming noise and very high sound insulation is then needed. Where shielding is not possible, separated constructions may be appropriate. Sound insulation between studios can be high enough for this to be needed. Control rooms must be well insulated from the studios with deep

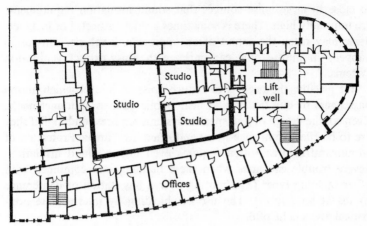

Fig. 78. Plan of Broadcasting House, London

double glazing, which is carefully detailed to achieve 45+ dB nominal. Studios must be protected by sound lock lobbies with well sealed solid-core doors. Where large access doors are needed, it is normally necessary to use very heavy purpose-designed doors with a carefully detailed sealing mechanism.

Circulation noise and impact noise should be limited by using absorptive ceilings/carpets in circulation areas and resilient floor coverings in any areas alongside or above studios.

Noise control inside the studio means attention to detail—to clocks and lights. The ventilation noise is a major item, particularly in television studios where lighting loads call for a large number of air changes. Air speeds must be kept down to e.g. 1 m/s at the discharge. Duct sizes must therefore be large, taking up a great deal of space, and very careful noise control is needed throughout the system. Noise control in studios is normally best dealt with by specialists.

Surgeries, Consulting Rooms and First-Aid Rooms

To avoid overhearing between consulting/treatment rooms and waiting areas, sound insulation must be high and background sound levels must not be too low. (See p. 175.) It is difficult to get sufficient insulation with a single door, if people are waiting immediately outside. A sound lock is useful if space allows.

Transport Buildings

Bus stations, airports and railway stations are examples of buildings so close to large noise sources that sound insulation requirements are unusually high. There is sometimes a safety aspect. For instance, noise may mask the sound of an approaching vehicle. This often happens in multi-storey car parks with noisy extract ventilation systems.

But usually, the need is to reduce noise to a level which allows passengers to buy tickets and hear public address announcements. Even so, the design of loudspeaker systems requires some skill if they are to be effective in such noisy and often 'live' areas. Protection of administrative areas often requires even higher sound insulation. Severe trouble can arise when these buildings are combined with other building types (e.g. housing over a bus station or conference rooms at an airport). The use of shielding by intermediate non-critical areas is helpful.

9

Sound Measurement and Calculation

In the first part of this chapter we describe briefly the apparatus used for making sound measurements, excluding the more complicated equipment which will only be employed by those who specialise in the subject. In the second part we describe the types of noise measurements that can be made, and give the various relationships and corrections that are useful with such measurements. Many of these relationships and corrections are only approximate, either because the conditions are such that precise calculations are not possible or because there is, at the present time, not sufficient practical knowledge.

The accuracy with which noise levels can be measured depends on the circumstances, but it is seldom that there is any need to give the results more accurately than to the nearest decibel. To be on the safe side, sound pressure level measurements should be rounded-off upwards and sound insulation measurements downwards.

APPARATUS

There are three basic parts involved in sound measurements: a microphone, an amplifier and an indicating device (e.g. a meter). Nowadays these are usually combined in one piece of apparatus, the sound level meter, and we will confine ourselves mainly to this instrument.

Modern equipment, of a good make, is accurate and reliable but still needs careful use and adequate maintenance. The first item in the chain—the microphone—will usually be a condenser or crystal one (their mode of operation need not concern us here), and will have been calibrated—in conjunction with the rest of the chain—by the manufacturer, i.e. a certain sound pressure level at the microphone will produce the appropriate reading on the meter. It should be emphasised that microphones are rather fragile, and if damaged, may not stop working altogether but may give out a smaller voltage (never greater) than their calibration. It is thus easy to get noise readings lower than they should be, and thus the microphone

205

sensitivity should be checked regularly. This may be done by using a standard noise source. A common device is a pistonphone, which is usually placed over the microphone and which contains a piston driven by a battery at a fixed frequency and at a fixed amplitude, so that the sound pressure produced in the enclosed volume containing the microphone and the piston is standard. Note that this device is not a complete check, because it does not check the response of the microphone at different frequencies. However, if modern microphones do go wrong it is usually over the whole frequency range, so the fault will show up.

Microphones used for noise and sound insulation measurements are, or should be, non-directional (except for very special purposes). However, because of their size these microphones unavoidably become directional at the higher frequencies; also the body of the sound level meter, and the body of the person holding the meter, will shade the higher frequencies from the microphone. Thus, usually it is better to 'point' the microphone towards the noise source to avoid this shading. Extension cables are available so that the microphone can be at some distance, say 1 m from the body of the sound level meter.

The lower limit of noise that microphones will measure down to is set by the unavoidable electrical noise of the circuit relative to the sensitivity of the microphone, but usually this lower limit is quite low enough for most practical purposes.

When microphones are used out of doors it is nearly always necessary to protect them from the wind, which otherwise will make a lot of noise owing to turbulence round the microphone. This is done by using silk or muslin or nylon stretched over a wire frame completely enclosing the microphone. The volume enclosed should be as large as possible, and for very severe conditions it may be necessary to enclose the first wind shield by a second one.

Microphones are vulnerable to damp, so must be kept dry when used out of doors. Rain shields are available, but they do have some effect on the response of the microphone, particularly at the higher frequencies.

The manufacturers' literature should give the values for the effects described above.

The range of sound pressures the ear will cope with (see p. 23), is from about 0.00003 Pa up to about 30 Pa, i.e. a factor of one million, so even if the lower limit (set by the electrical noise) is, say, 0.0003 Pa, a measuring amplifier still has to deal with voltages in the ratio of 1 to 100,000. Microphones themselves are designed to respond linearly over this range of pressures (i.e. if the pressure increases by a factor of two, so does the voltage output) and in fact most common microphones respond linearly up to about 140 dB. Above that, they will tend to give distorted results, and may be permanently damaged. However, the amplifier following a microphone will not usually be able to handle such a big voltage range, and obviously a meter scale could not show it. It is usual to put calibrated attenuators in the amplifier circuit which are adjusted by hand for the particular noise being measured so that the meter reading is convenient. The meter scale is calibrated in decibels, so the total noise level will be given by the sum of the attenuator setting and the meter reading.

The third item in the measuring chain of the sound level meter is the meter itself. The audio-frequency voltages must be rectified in some way before they will operate a meter. It is better, for various reasons, if the meter circuit is designed to measure the root-mean-square value, but often the mean rectified level is measured. Rarely, the peak level is measured but it should be realised that this is the peak level of the wave-form and not the peak—or maximum—value reached by a fluctuating noise.

All meters have two standardised responses—'fast' and 'slow'. Their meanings are obvious; the slow response is useful to get an average answer for noises which are fluctuating rapidly (such as the noise from a pneumatic hammer). If the noise is changing a lot with time, e.g. a passing aircraft, then more elaborate means than a simple meter are needed to measure how the noise level varies with time (see below).

Some sound level meters have a switch so that they indicate the maximum value of a transient noise, i.e. the meter goes up to the maximum noise level reached and stays there—giving adequate time for an accurate reading—until the reset button is pressed. (Such maximum values should not be confused with the peak values of a wave-form, see above.)

Sound level meters have built into them 'weighting networks', i.e. electrical circuits arranged to attenuate different frequencies by different amounts. The internationally standardised networks, known as A, B and C, are shown in Fig. 58. Fig. 5 (p. 26) shows that the ear's sensitivity varies with frequency, and the original idea of the sound level meter's weighting networks was to simulate the response of the ear, so as to give a loudness reading in phons (see

p. 26). Thus the A-weighting was used when the sound pressure level was below 55 dB, because the response was something like the shape of the A curve, i.e. less sensitive at the lower and highest frequencies. Similarly, the B response was to be used for sound pressure levels between 55 and 85 dB, and the C-weighting for higher levels. However, it was found that this simple idea did not work, i.e. did not give reliable loudness readings, but it was subsequently found that for many common noises an even simpler idea did work, and that was to use the A-weighting network for all levels of noise. The answer obtained was not then numerically in phons, i.e. a reading on the sound level meter of say, 80 dB with the A-weighting network switched in did not indicate that the noise had a loudness of 80 phons, but it was found that there was a reliable and constant relationship between the readings obtained with the A-weighting (and known as dBA) and the loudness of a noise. As explained on p. 26, true phons can only be obtained under laboratory conditions, and so their use in practical situations has died out, the answers and criteria usually being given directly in dBA (or very occasionally in dBB, i.e. with the B network in, or dBC, or PNdB, see p. 225). When the answers are given in this form (e.g. dBA) they are referred to as the 'sound level', compared with answers in, say, octave bands, which are referred to as 'sound pressure levels'.

If the noise is varying greatly with time, e.g. as does the noise from road traffic, then while the maximum value can probably be read off a meter without too much difficulty, it will not be possible to get any measurement of how the noise level varies with time. For these cases some form of level recording instrument is desirable. An ordinary pen-recording milliameter could be used in place of the meter, but this will have a linear scale. That is to say, if its full-scale deflection is, say, 10 and the minimum deflection that can be read is 0·5 this factor of 20 is only 26 dB, which may not be a big enough range. What is more, the answers will be wanted in dB above 0.00002 Pa and all the linear readings on the pen recorder will have to be converted into logarithmic units. A much more useful (but more elaborate) recorder to use is what is called a 'high-speed level recorder'. This type of recorder traces out on the recording paper the logarithm of the voltage applied to the instrument. Thus it can be calibrated— in connection with the sound level meter—to read directly in decibels. Its range can be selected by using interchangeable input potentiometers, the most commonly used range being 50 dB, but 25 dB or 75 dB ranges are also available. The speed at which the paper travels through the recorder can also be selected at will, from a few milli-metres an hour (for use when noise levels over a long period are required) up to 100 mm a second (when noises which are changing in level very rapidly are being measured). This type of recorder will respond very quickly if desired—at a rate of up to about 1000 dB per second—but its response rate can be slowed down as required.

For more detailed information about the frequency content of a noise, such as is necessary, for example, when assessing the accept-ability of ventilation-plant noise, a frequency analysis can be made using electrical filters introduced into the measuring equipment, so that only the noise in one frequency band reaches the meter or level recorder at one time. Such band-pass filters are of two basic types. The first is the constant bandwidth type, and a common bandwidth is 5 Hz (although other values are available). This means that, for example, when the filter is set to 100 Hz only the noise between 97·5 and 102·5 Hz is measured. If it is set to 1000 Hz then the band passed is 997·5 to 1002.5 Hz. It is obvious that this type of filter gives a very detailed analysis of a noise, in fact often too detailed for practical use.

The other class of analyser has a constant-percentage bandwidth (i.e. a bandwidth which is a constant percentage of the centre fre-quency of the band). They are also known as proportional analysers (i.e. the bandwidth is proportional to the centre frequency). These analysers can pass either what are known as narrow bands or broad bands. A common example of a narrow-band analyser is one with a bandwidth of 3%, that is to say it passes frequencies lying within 1·5% of the frequency it is set at. Thus, for example, at 100 Hz it would pass 98·5 to 101·5 Hz, and at 1000 Hz it would pass 985 to 1015 Hz.

For nearly all the purposes of this book a broad-band constant-percentage analyser is of most interest, and the two sorts we are con-cerned with are one-third-octave filters and one-octave filters (many sound level meters include facilities for adding such filters). One-third-octave filters pass, of course, bands of noise one-third of an octave wide, and the centre frequencies of successive bands are thus one third-octave apart. The results of octave or one-third-octave analysers are usually presented as a graph with each band represented by a point at the centre frequency. The standardised centre frequen-cies for one-third octave and octave filters are given in Table V (see also Fig. 2).

Table V Standardised centre frequencies for one-third octave and octave filters

One-third Octave centre frequencies (Hz)	Octave bands centre frequencies (Hz)
(Lower than 50 if necessary)	31·5
50	
63	63
80	
100	
125	125
160	
200	
250	250
315	
400	
500	500
630	
800	
1000	1000
1250	
1600	
2000	2000
2500	
3150	
4000	4000
5000	
6300	
8000	8000
10,000	

Of course when such filters are used in conjunction with sound meters, none of the weighting networks is used, the sound level meter being set to linear, i.e. equal response to all frequencies.

Occasionally, the exact wave-form of a noise is needed, and then the output from the amplifier is connected to a cathode-ray oscilloscope and the displayed wave-form observed or photographed. This technique is limited usually to the investigation of the sources of machine noise where it may be desirable to relate the wave-form with a particular motion of the machine. It should be noted that for this type of analysis (unlike the other noise measurements described below) the phase changes caused by the microphone and measuring apparatus are important.

If a frequency analysis of a transient noise is required, then obviously it will be necessary to repeat the measurement for each frequency band. Thus for octave-band analysis eight measurements would have to be made. A more convenient way is to record the noise on magnetic tape and replay it subsequently through the octave bands on to a level recorder. This order of technique is rather beyond the scope of this book, but a few points to be watched will be mentioned. An electrical calibration in terms of the calibration of the microphone should be recorded on the tape at the start and finish of each set of measurements. The gain control of the tape recorder (which usually controls both the recording level and the play-back level) should be calibrated in decibels. Good-quality tapes do not vary in sensitivity along their length by more than ± 1 dB. If the electrical calibration is recorded as a glide tone rather than as discrete frequencies, the response of the filter will be drawn out automatically on the level recorder during replay and will serve as a check on the operation of the circuit. The recording level on the tape should be as high as possible while not overloading it; this is because the signal-to-noise ratio of the tape recorder will probably not be as good as might be desirable for recording a noise which is changing a lot. The tape machine should be reasonably free from 'wow' and 'flutter', but for noise measurements this is not as critical as it is for speech or music. The speed of the machine must be constant.

MEASUREMENT AND CALCULATION

Adding Sound Pressure Levels

When it is necessary to sum two random noises, the intensities, not the pressures, must be added. For example, if two equal noises are added, the intensity will be doubled. But, as the pressure is proportional to the square root of the intensity, the pressure will go up by a factor of $\sqrt{2}$, in terms of decibels, an increase of 3 dB. Fig. 79 allows a simple means of adding pressures in decibels; the difference between the levels to be added is plotted against the number of dB to be added to the smaller level to achieve the sum. For example, to add pressure levels of 80 and 85 dB, the difference is 5 dB which,

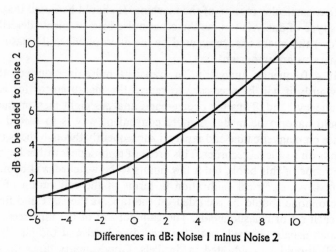

Fig. 79. Decibel addition

from Fig. 79, indicates that 6 dB should be added to the smaller (i.e. to the 80 dB). So the total level is 86 dB.

Similar considerations apply when, as happens in some instances, it may be desirable to calculate, from an octave analysis, the total sound pressure over the whole frequency range. To do this, the intensities in each octave band must be added. For example the octave band levels of Fig. 80 are given in Table VI, and the intensities in each octave band relative to some convenient reference (in

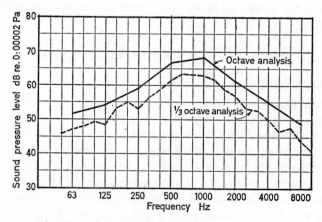

Fig. 80. Analysis of noise in a canteen

212

this case 60 dB) are calculated from intensity ratios on the dB Table on p. 289. The sum of all these relative intensities is 14·15 and this is 11·5 dB above the reference of 60 dB. The total level over the whole frequency range is therefore $60 + 11·5 = 71·5$ dB. This addition could also be done using Fig. 79 several times. For accurate results, the lower values should be combined first.

Table VI Addition of sound pressures—example

Octave band centre (Hz)	Sound pressure dB re 2×10^{-5} Pa	Intensity relative to a level of 60 dB
63	52	0·15
125	54	0·25
250	59	0·80
500	67	5·0
1000	68	6·3
2000	61	1·25
4000	55	0·30
8000	49	0·10
		Sum 14·15

Effect of Bandwidth

For all noises except those consisting of a few discrete frequencies, there will obviously be more energy in a wide frequency range than in a narrow range. Therefore in referring to measurements or calculations, it is essential to say within what frequency range sound pressures are being considered, e.g. in octaves or one-third octaves. Occasionally, we need to convert results in narrow bands to results in broader bands. The conversion is made by adding $10 \log \triangle f_1 / \triangle f_2$ (in dB) to the measured results, where $\triangle f_1$ is the broader bandwidth and $\triangle f_2$ the narrower range. For example, we may need to convert one-third-octave data to octave-band data. In this case, $\triangle f_1 = 3 \times \triangle f_2$. Therefore $10 \log \triangle f_1 / \triangle f_2 = 10 \log 3 = 5$ dB. So 5 dB must be added to one-third octave-band measurements to convert them to equivalent octave bands. This applies to the whole frequency range, because $\triangle f_1 / \triangle f_2$ remains the same throughout in any change from one-third-octave to octave. This is illustrated in Fig. 80, which shows the noise in a canteen measured in octave bands and one-third-octave bands. Of course, the one-third-octave bands show more detail, and in general are 5 dB lower than the octave-band results.

Noise out of Doors

(a) Noise Sources

For individual stationary noise sources, e.g. machinery, recommended methods of obtaining data about the source include:

1. Measurements in the absence of reflections.
2. Measurements in a reverberation chamber.
3. Measurements in semi-reverberant conditions.

Measurements made in free field conditions (referring to anechoic or outdoor locations) have the advantage of giving more information about the direction in which the sound leaves the machine than other methods. But the total power output is complicated to measure unless the energy is collected inside a hard room (method 2 above). An assessment of the total output of a machine in free field conditions can be made from sound pressure measurements at many points around and over the top of the machine, either at approximately 1 metre from the surface of the machine (sometimes referred to as the 'prescribed surface' method) or over the surface area of a hemi-sphere around the source. Measurements in semi-reverberant con-ditions are also possible if room conditions are fully known and/or a comparison is made with a known noise source.

More often, the sound pressure level at a given distance in the direction of interest is already known or can be measured. Single measurements should not be made too close to the source (i.e. at a distance less than $1\frac{1}{2} \times$ the largest dimension of the exposed face of the source) since there will often be very large variations resulting from mutual reinforcement and cancellation of sound from the various parts of the source (in the near field, see p. 128). Measure-ments must not be made so far away from the source (> 30 metres) that other forms of attenuation e.g. in the air or at the ground, have a strong influence on the results. Even if the directivity for the direction of interest is known, it is wise to check on the variation in directivity to allow for atmospheric effects or reflections redirecting the sound towards the listening position.

(b) Effect of Distance

The first factor affecting reduction of noise level with distance from a source is the spread of the energy. For a small source, we see from the geometry of the sphere that the energy per unit area at a point is inversely proportional to the square of the distance from the source, i.e.

$$\text{Intensity} \propto \frac{1}{\text{distance}^2}.$$

In terms of decibels, this means that for every doubling of distance, we can expect a reduction of 6 dB. This is the inverse square law

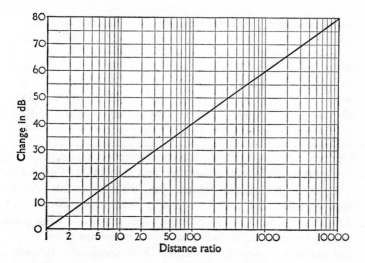

Fig. 81. Reduction due to the inverse square law

(see Fig. 81). If there are reflecting surfaces nearby, the reduction will be less. However, out of doors, if separate allowance is made for the effect of the ground, the inverse square law provides a useful basis for estimating noise levels for small sources. For larger noise sources, the reduction in sound pressure levels with distance will be less than predicted by the inverse square law. For distances from the source less than one-third of the largest dimension of the source, e.g. for a train of 90 metres length, the inverse square law is not likely to apply until the receiving position is more than 30 metres from the train. Care must be taken not to extrapolate measured data (taken near a large source) using the inverse square law.

The second factor is the molecular absorption of sound in air. This is only important at high frequencies and varies with temperature and humidity. Measured values which have been reported vary somewhat. But the data in Table VII provide a guide to the attenuation in air at different frequencies for varying conditions of temperature and humidity.

Table VII Molecular absorption of sound in air

	Attenuation dB/1000m					
	21° C			2° C		
Octave band centre (Hz)	Relative humidity			Relative humidity		
	40%	60%	80%	40%	60%	80%
1000	3	3	3	10	6	0
2000	13	6	6	33	16	3
4000	33	16	16	49	49	33
8000	130	82	49	82	130	82

Where sound is propagated in a vertical direction e.g. from aircraft to the ground, the inverse square law and absorption in air are the two important attenuation factors. But, with the source near the ground or where the path of sound is nearer to horizontal (i.e. striking the ground at a shallow angle), temperature or wind gradients and ground attenuation become most important.

Unfortunately, propagation under these conditions is such a complex phenomenon that accurate prediction is not possible. One of the many complications is that attenuation will often depend on the absolute distance and cannot be expressed as so many dB per 1000 m.

The effect of a wind gradient is to reduce the intensity of the sound upwind and perhaps to increase it downwind. The effect of a temperature gradient during the daytime (when usually temperature decreases with height) is to decrease the sound level at a distance from the source, depending on the magnitude of the temperature gradient and on the height of the source and the receiver above the ground. Conversely, at night, when the temperature increases with height, the sound level may be increased at a distance. Variation due to wind and temperature effects are typical over a range from 5 dB less attenuation than predicted by the inverse square law (when wind and temperature are helping sound propagation) to 20–25 dB more attenuation (when they are inhibiting propagation). Turbulence will scatter sound and may well feed it back into a shadow zone left by temperature or wind gradients or screening. As a result, excess attenuation tends to be limited to 20–25 dB in these cases. Fog and rain will also modify attenuation with distance. But any change in absorption may be counteracted by the absence of large wind or temperature effects, so that results are difficult to predict accurately. It is interesting to note that during rain, the background noise level will often rise appreciably, making it more difficult to hear other noise. But in almost all cases, benefits from weather conditions cannot be allowed for in calculations, and these are second-order effects.

Ground attenuation for nearly horizontal propagation is yet another complex phenomenon and is strongest when the source and

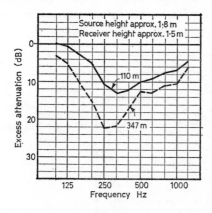

Fig. 82

An example of attenuation over grassland

receiver are close to the ground. A change of phase on reflection at the absorbing ground surface results in some cancellation of noise over a band of frequencies (usually between 200 Hz and 600 Hz), resulting in excess attenuation over this range, sometimes as much as 20–25 dB. Apart from this the attenuation over soft ground varies significantly from that predicted by the inverse square law.

A wide range of values for excess attenuation due to propagation over soft ground (e.g. grassland or ploughed fields) have been reported. They are of the type illustrated in Fig. 82, which gives one set of results based on a source height of approximately 1·8 metres and a receiver height of approximately 1·5 metres, for two distances from the source. The variations are so great depending on the exact nature of the ground that it is impossible to give a general and useful guide. Estimation of propagation of road traffic noise (a line source rather than a point source) over grassland is given in Fig. 88. Over hard ground the use of the inverse square law is appropriate for small sources. For line sources the attenuation is less (see Table IX).

(c) Screening

To reduce the transmission of sound out of doors, some obstacle such as an earth bank, a wall or building can be interposed between the source and receiving position. The reduction obtained by such a screen is limited by the passage of sound over the top and around the side of it. Further, if any appreciable ground attenuation is present, it may be lessened (and the effective reduction by the screen is therefore also lessened) because the path of the sound over the wall is further away from the ground than the direct path would have been. If this is the case, calculations must allow a reduction of ground attenuation. As a guide, we can use the attenuation over hard ground as a guide to ground effects in this situation.

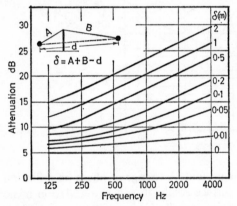

Fig. 83
Attenuation of noise from a point source by a very long screen

Attenuation due to a screen is related to how much further the sound is forced to travel in going round the screen compared with the direct path. From Fig. 83 the important quantity is the extra distance $\delta = A + B$—the direct distance. For a very long barrier (long enough for the sound passing round the ends of the barrier to produce a negligible contribution at the receiving point) the attenuation is given in Fig. 83. It will be seen that even with a direct line of sight between source and receiver, some attenuation occurs, owing to the close proximity of the screen. An approximate estimate of attenuation by a smaller screen can be made by carrying out the screening calculation three times and adding the results in reverse.

218

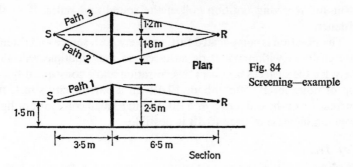

Fig. 84
Screening—example

Example: Referring to Fig. 84, calculate the attenuation at 500 Hz at the listening position R due to the presence of the screen shown.

(The path difference) for Path 1 $= 0.2$ m. Attenuation $= 11$ dB.
                         Path 2 $= 0.7$ m. Attenuation $= 16$ dB.
                         Path 3 $= 0.3$ m. Attenuation $= 12$ dB.
Combining $- 11$ and $- 16$, we get $- 10$.
Combining $- 12$ and $- 10$, we get $- 8$.
No ground effect applies before or after the screening.
Therefore the total screening effect is approximately $- 8$ dB.

Note that, were the screen of the same height and very long, the attenuation value would be as for path 1, i.e. 11 dB. Similarly, if the screen were very high, the performance would be determined by paths 2 and 3, which, when combined also happen to give attenuation of 11 dB. Therefore in our example, half the energy is passing over the top of the screen and half around the sides.

The wind can affect screening to the extent that, at high frequencies, attenuation by screens can be reduced by 20 dB+ at wind speeds greater than 30 km/hour.

(d) Reflections near the Source or Listener

Reflecting surfaces near the source effectively make it more directional. If the sound is reflected towards the listener from a single plane surface near the source, an increase of 2 or 3 dB may be expected. If two reflecting surfaces are involved an increase of 4 or 5 dB is likely. The more the source is enclosed by reflecting surfaces which redirect the sound towards the listener, the greater the directivity of the source and the higher the noise level at the receiving position. Similar increased noise levels occur if reflecting surfaces

219

near the receiving position collect the sound and reflect it to the listener.

The situation is complicated further where both source and listener are enclosed within reflecting surfaces, so that multiple reflection between the surfaces produces reverberation and a consequent build-up of noise levels, often about 10 dB. In a narrow city street, the increase is of the order of 5 or 6 dB. But in a small courtyard or light well, an increase of over 10 dB is possible.

(e) Trees

If sound travels any appreciable distance through belts of trees, then some additional attenuation will be introduced. Table VIII gives some approximate attenuation figures for woods or forests of average density—in leaf or as bare trees.

Table VIII Excess attenuation for sound propagation through trees (dB per 100 metres).

Octave band centre (Hz)	Average woodland with foliage	Bare trees (average density)
63	1	0
125	2	0
250	3	0
500	3·5	1
1000	4	1
2000	5·5	2
4000	8	3
8000	10·5	4

Road-Traffic Noise

Calculation of road-traffic noise is usually based on L_{eq} or L_{10} values in dBA (see p. 162). For direct correlation with subjective reactions, the use of L_{10} (18-hour) may be appropriate. Often an L_{10} (1 hour) applying to the busiest period affecting the building of interest is more useful.

Estimation of noise levels from free-flowing road traffic can be quite accurate, if a large number of variables are considered. Where estimates are needed to satisfy legal cases or for decisions involving large expense, the reader is referred to the detailed procedures in the current official publications. However, we will look at the main

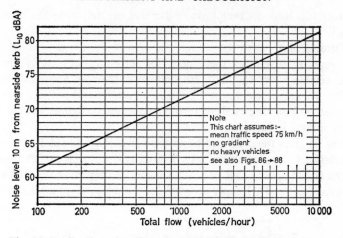

Fig. 85. Predicted road-traffic noise level—L_{10} over one hour, at 10 m from the nearside kerb, for free-flowing traffic

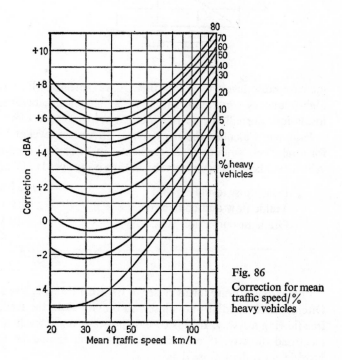

Fig. 86

Correction for mean traffic speed/% heavy vehicles

factors involved, so that broad estimates can be made for simple cases.

Dealing firstly with the source, the noise output is dependent on the quantity of vehicles, their speed, the percentage of heavy vehicles (i.e. weighing over 1500 kg), the gradient of the road and its surface finish. We can calculate the noise level at a reference position (often 10 metres from the nearest kerb) by combining figures related to each of these influences.

Fig. 85 gives the predicted basic noise level (L_{10}dBA) at 10 metres, in relation to traffic flow and Fig. 86 gives corrections for mean traffic speed and the percentage of heavy vehicles. A correction for gradient is given in Fig. 87. Deeply grooved (> 5 mm) surfaces are likely to

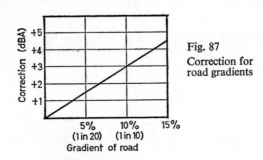

Fig. 87
Correction for
road gradients

increase noise levels by 3 to 4dBA. A wet surface will increase the high-frequency noise. But this is rarely significant, because sound insulation is usually better than needed at high frequencies.

Example: Calculate the L_{10} (1 hour) dBA noise level at 10 metres for a relatively smooth road (gradient 1 in 20 or 5%) carrying 1200 vehicles per hour (20% heavy vehicles) at a mean speed of 60 km/hour:

Basic noise level (Fig. 85)	72 dBA
Traffic flow and content (Fig. 86)	+ 2.5 dBA
Gradient correction (Fig. 87)	+ 1.5 dBA
	76 dBA

Prediction of noise levels closer to the road traffic is less accurate. Often, where buildings are closer, as on city sites, the traffic is not free-flowing anyway. In this case, if possible, noise levels should be measured directly. If not, a similar situation should be found to provide equivalent noise data.

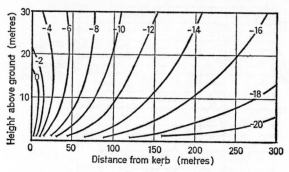

Fig. 88. Correction for propagation of road-traffic noise over grassland

Often, buildings are more than 10 metres from the road and we need a prediction method for reduction of noise level with distance from the source. Where the ground between is hard, attenuation is as shown in Table IX. For soft ground, e.g. grassland, Fig. 88 should be used.

Table IX Approximate attenuation from roadway over hard ground for free-flowing traffic.

Distance from nearside kerb (m)	Correction (dBA)
10	0
20	− 2·5
50	− 5·5
100	− 8·5
150	− 10·5
200	− 11·5
250	− 12·5
300	− 13·5

There are, of course, numerous variations of road geometry. Detailed calculations must take account of the build-up of noise in cuttings or between buildings, the effects of curvature of the road, variable gradients and screening. Fig. 89 gives a method of assessing the simple case of screening by a very long barrier in terms of dBA. As we have said, the lower part of the sound field, which is most significant from the point of view of ground effects, is screened. Therefore distance corrections used with screening tend to be those

223

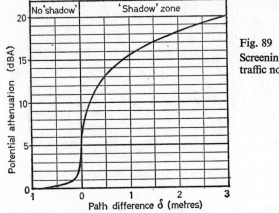

Fig. 89
Screening of road-traffic noise

in Table IX, unless the screen is very low (< 0·2 metres) and the source receiver distance is very long(> 150 metres).

If dual carriageways are separated by more than 3·5 metres, calculations should be carried out treating each carriageway separately and adding the estimated levels for each.

Where the final level required is 1 to 2 metres from a building façade (as a basis for sound insulation calculations), the effect of a local build-up due to the presence of the façade may need to be accounted for (usually as a correction of + 2.5 dBA).

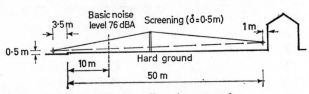

Fig. 90. Estimation of road-traffic noise—example

Example: Extending the previous example (giving a basic noise level at 10 m of 76 dBA), calculate the noise level at 50 metres from the kerb, over hard ground, with screening as shown in Fig. 90, at a receiving position 1 metre in front of a building façade:

Basic noise level	76 dBA
Screening ($\delta = 0.5$ m)	− 13 dBA
Distance correction (from Table IX)	− 5·5 dBA
Reflection at façade	+ 2·5 dBA
Resultant level	60 dBA

Note that with soft ground, the result would not change, since with screening, the distance correction remains as for hard ground.

Calculations become more complicated where a detailed sound spectrum is needed. More often than not, the calculation error is such with a frequency breakdown that estimation using dBA gives equally good results. Nevertheless it is wise to consider the spectrum.

Railway Noise

Railway noise is by no means as well documented as road-traffic noise. There are few new railway lines and there appears to be considerable tolerance of the noise from existing ones. The passage of a train is usually very much more a single event and therefore a different form of intrusion.

The main influences on railway noise are the number of passes, the form of traction, speed, loadings and track details. In almost all cases, since the railway usually exists already, site surveys provide the most accurate data and insulation calculations (see p. 239) should be based on reducing peak noise levels to match the criteria on p. 172. In most cases, calculations based on dBA are accurate enough. But for consistent forms of traffic, it may be appropriate to make calibrated tape recordings on site for spectrum analysis later. There is a danger that calculations based on dBA take too little account of low-frequency components.

A survey of railway noise should include a check on vibration levels, where buildings or their occupants could be affected.

Aircraft Noise

In dealing with aircraft noise, we are usually concerned to find the insulation needed to protect buildings in and around airports or under low flight paths. In some cases, e.g. where new airports are being developed, estimates have to be made either from data obtained at existing airports, catering for similar traffic or from published data on individual aircraft types (this data is available for the more common types of aircraft, usually as PNdB contours). PNdB values can be related to dBA values approximately by the formula:

$$PNdB \simeq dBA + 14.$$

Information on detailed spectrum content is usually more difficult to obtain.

If the site is affected by existing air traffic, a direct survey of noise

H   225

levels is useful, provided that conditions of measurement are representative, i.e. allowance is made for changes in flight paths from day to day, atmospheric conditions and the variety of aircraft involved. Noise exposure on a given site is likely to vary considerably from day to day, e.g. being subject to landing or take-off, subject to wind direction. Traffic may vary from the smallest private plane to the largest passenger or freight aircraft. Aircraft authorities or traffic control departments can be very helpful in providing records of flight timetables, which often give more detail than is needed and allow a check on whether the survey data is representative. For major airports, NNI contours (see p. 164) are often available.

A wide range of units has been developed to describe aircraft noise. But many, such as NNI are not easily applied directly to sound insulation calculations. Again, as with railway noise, a good guide can be given by applying the criteria on p. 171. Measurement of peak dBA or recordings analysed in octave bands provide a sensible basis for calculations. Both for aircraft and railway noise, as well as other transient sources, peak noise levels must be measured to a specified meter response speed (usually 'fast' response). Where costly decisions are to be based on surveys or calculation, specialist advice should be sought.

The calculation of required insulation, based on road, rail or aircraft noise levels at 1 m from buildings is discussed on p. 239.

Noise Indoors

The simplest measurement indoors (as for out of doors) is that of a noise level at a given position or positions. An example is when we need to know the noise levels to which an operator of a machine is exposed. The microphone is placed at the operator's head position and the sound pressure level measured.

As explained in Chapters 2 and 6, when a noise source is operating in a room, the direct sound will predominate at positions close to the source and the reverberant sound at positions away from the source. If the source is non-directional, then the direct sound level will equal the reverberant sound level at a distance, r from the source, given by:

$$r = \frac{\sqrt{\dfrac{A}{1 - \bar{\alpha}}}}{7} \text{ metres}$$

where A = total absorption units in room (m²) and $\bar{\alpha}$ is the mean

226

absorption coefficient (see p. 40). In a relatively 'live' room, $\bar{\alpha}$ will be small compared with 1 and the formula becomes:

$$r = \frac{\sqrt{A}}{7} \text{ metres.}$$

If, as is more usual, the source is directional, the conditions are much more complicated and calculations in detail cannot be dealt with here. Nevertheless, measurements at different distances from a source will quickly give an indication of the distance over which the direct sound predominates (ref. Fig. 45, p. 133). It must be remembered that when absorption is added to a room, thereby lowering the reverberant sound level, the direct sound will predominate over a larger distance.

Unless we are checking the levels at given positions, as described above, the reverberant sound is generally of more interest. This is because the reverberant sound level is needed either (a) for estimating the reduction in the sound pressure level that would result from the introduction of extra sound absorption into the room, or (b) for many calculations related to sound insulation.

But the reverberant sound level in a room will not be uniform, unless the room is so large and irregularly shaped that the sound field is completely diffuse—this is not always the case in practice. The sound pressure level must be measured at several positions to obtain a representative value. The measurement positions should be chosen so that there is no doubt that it is the reverberant field that is being sampled, but should not be too close to any of the room surfaces. This is because, depending on the nature of the surface, the pressure close to the surface (i.e. less than half a wavelength away) will be higher than the reverberant sound pressure. In practice, where specifications for background noise limits are given, they are typically measured at 1·5 metres height above the floor and at least 1 metre from the boundaries. In small rooms, there may be difficulty in finding appropriate measurement positions. If this is the case, it is likely that we could not be sure that we are measuring the reverberant field anyway.

Absorption in Rooms

The absorption coefficient of a material refers to the proportion of incident sound which is absorbed by the material. For example, an absorption coefficient of 0·3 indicates that 30% of the sound incident on the surface will be absorbed by it, the other 70% being reflected.

The total absorption of a surface is given by the absorption coefficient multiplied by the area i.e. $S \times \alpha$. If our surface is 20 m² in area, $S = 20$, $\alpha = 0.3$. The total absorption is therefore 6 m² units.

The absorption coefficients of materials vary with frequency. Appendix A gives values for some common materials. Absorption by room contents may be difficult to assess on the basis of surface area, e.g. people, furniture. In practice, it is more realistic to deal with these items on the basis of a given number of absorption units per person or per seat. These units (also given in Appendix A) are then multiplied by the number of items: e.g. at 500 Hz, the absorption of 50 people seated in relatively hard seats will be:

$$50 \times 0.4 \text{ m}^2 \text{ units, i.e. } 20 \text{ m}^2 \text{ units.}$$

However, experience has shown that for large audience areas, it can be more accurate to deal with area of audience with specific coefficients related to this approach, as with the primary room surfaces (see Appendix A).

At frequencies of 1000 Hz and above, the absorption in the air becomes important, and depends on the temperature and humidity. For typical room temperature and humidity, the absorption in m² units (for convenience in adding to the remaining absorption) is $x \times$ the volume of air (m³), where x is the coefficient for air given in Appendix A.

We are now in a position to calculate the total absorption of a room in line with the principles set out on p. 42. Let us take the example of a room of volume 62.5 m³ containing five people seated on wooden chairs. The ceiling area is 20 m² of 12 mm plasterboard under joists, the floor 20 m² medium pile carpet on closed-cell foam underlay, on concrete, the walls 50 m² of plastered brickwork and 10 m² of window. The amounts of absorption in the 500 Hz octave band are:

	Sabins (m²)
Plasterboard 20 m² @ coeff. of 0.1	2
Carpet 20 m² @ coeff. of 0.3	6
Plaster on brick 50 m² @ coeff. of 0.02	1
Windows 10 m² @ coeff. of 0.1	1
Five people @ 0.4 sabins each	2
	—
Total absorption	12
	—

In this particular octave band (around 500 Hz), the air absorption is negligible, but if the absorption in, say, the octave band around 4000 Hz were needed, the amount $x \times V$ would be added, where x is the coefficient for air absorption and V is the volume of air (m³):

$$\text{i.e. } 0·02 \times 62·5 \approx 1·3 \text{ sabins.}$$

The intensity of the reverberant sound in a room is inversely proportional to the total absorption present, so that the change in reverberant sound pressure level, in decibels, due to a change in the total absorption present can be found from the intensity relationships in Appendix E. Doubling the absorption will reduce the level by 3 dB and quadrupling absorption by 6 dB, and so on.

If, for example, the reverberant sound pressure level in the room described above is 80 dB in the 500 Hz octave band, and the ceiling construction is changed to give very high absorption (an absorption coefficient of, say, 0·9), this will produce an increase of 16 sabins. The total absorption has therefore gone up from 12 sabins to 28 sabins. The ratio 28:12 equals approximately 2·3 which, from Appendix E equals an intensity ratio of about 4 dB. The reverberant sound pressure level will therefore be reduced to 80 − 4, i.e. 76 dB.

The amount of absorption in a room can be measured by measuring the reverberation time, as follows.

Reverberation Time

As we have said in Chapter 2, the reverberation time of a room is defined as the time taken for the sound pressure level in that room to drop by 60 dB, after the sound source is stopped. Since it is often awkward to have to produce noise levels as much as 60 dB above the background noise level, the value of reverberation time is usually found by measuring the gradient of the decay of the sound and projecting this to find a value over the full 60 dB. A loud noise source is used and as the noise output is stopped suddenly, the decay is recorded on a high-speed level recorder. (Alternatively, tape recordings are made of the decay of the sound for subsequent replay into a pen level recorder.)

The source should preferably be loudspeakers which are arranged to be essentially non-directional, fed with random noise in one-third-octave bands. But in rooms where the reverberation time is likely to be longer than about 1·5 seconds at mid-frequencies, impulsive noise sources, e.g. a pistol firing blanks, may be more convenient.

But care must be taken to avoid using impulsive sound sources which 'ring' after being used.

The noise level should be more than 40 dB above the background noise level. Microphone positions should be well clear of the zone where direct sound predominates and a minimum of three measurement positions should be used, preferably more and certainly more in large rooms. At least two samples should be collected at each microphone position.

Since reverberation time varies with frequency, filters will normally be interposed between the microphone and the pen level recorder. It is most important that pen and paper speeds on the level recorder are accurate. Where tape recordings are made, tape speeds must be consistent. When analysing the decay traces, the part of the trace corresponding to 5 to 35 dB below the initial level should be used to generate a straight line, giving the slope to be used to obtain the reverberation time. If a double slope occurs (see p. 46), the range of reverberation time values should be stated. But curved decay slopes are almost impossible to evaluate. The reverberation time can then be obtained by extending the slope to achieve a decay of 60 dB and reading off the time or, more usually, by means of a protractor designed for the purpose. Methods for measuring reverberation times in auditoria are usually standardised in official publications.

Having measured reverberation time, the total absorption (A) in m² can be obtained from Sabine's formula (see p. 42):

$$RT = \frac{0 \cdot 16 \times V}{A} \text{ seconds}$$

where V is the room volume in m³. Of course, at high frequencies, A will include air absorption. The measurement of reverberation time is a more accurate way of getting the total absorption in a room than calculating as described previously. If we take our previous example again, i.e. a room of volume 62·5 m³, with surfaces as previously described, but without contents and we measure reverberation time at 500 Hz of 1·1 seconds, we can calculate the total absorption to be $\frac{0 \cdot 16 \times 62 \cdot 5}{1 \cdot 1} = 9 \cdot 1$ sabins, compared with 10 sabins estimated before. If, however, we wish to estimate reverberation time from the formula, we use the estimate of absorption from p. 228. For the empty room:

$$RT = \frac{0.16 \times 62.5}{10} = 1 \text{ second.}$$

For the occupied room:

$$RT = \frac{0.16 \times 62.5}{12} = 0.85 \text{ seconds.}$$

But, for very absorbent rooms ($\bar{\alpha} > 0.25$), we must use the Eyring formula:

$$RT = \frac{0.16 \times V}{S(\log_e(1 - \bar{\alpha})) + xV} \text{seconds.}$$

If, therefore, in our example our room were to have three times as much absorption at 500 Hz when empty, i.e. 30 sabins,

$$\bar{\alpha} = \frac{30}{100} = 0.3.$$

From the Sabine formula:

$$RT = \frac{0.16 \times 62.5}{30} = 0.33 \text{ seconds.}$$

From the Eyring formula, using Fig. 21:

$$RT = \frac{0.16 \times 62.5}{0.36 \times 100} = 0.28 \text{ seconds.}$$

In estimating reverberation times, it is unrealistic to expect results better than $\pm 10\%$ and in complex cases, $\pm 20\%$ is quite usual. Results are not normally sufficiently accurate for the use of two places of decimals. In practice, the second place of decimals is rounded up or down to the nearest 0·05 seconds, from our example 0·33→0·35 and 0·28→0·30.

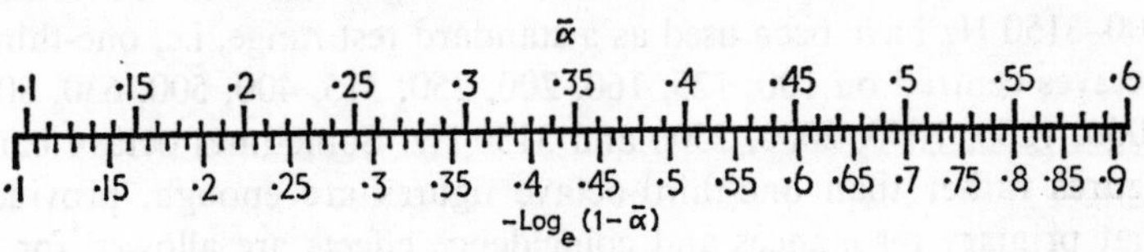

Fig. 91. Conversion of $\bar{\alpha} \rightarrow \log_e (1 - \bar{\alpha})$

From the Eyring formula, using Fig. 91, it will be clear that, unless rooms are extremely 'dead', the greater accuracy of the Eyring formula is not justified.

231

Airborne Sound Insulation

We will now refer to sound originating in the air, passing into the structure and being re-radiated by the structure into another space.

The sound transmission coefficient $\mathscr{T}$ of a partition (e.g. a wall, floor, roof or window) is the ratio of the sound energy transmitted through it to the sound energy incident on it. In decibels, the Sound Reduction Index (SRI) is $10 \log \frac{1}{\mathscr{T}}$. For example, if a partition re-radiates one five-hundredth of the energy incident on the other side, $\mathscr{T}$ is 0·002 and the SRI is 27 dB.

The airborne sound transmission varies with frequency and it is therefore necessary to measure it over a frequency range. This range should cover all frequencies likely to be important for the case at hand, but there are practical limitations. The lower frequency limit is set by the fact that the wavelength of the sound may be the same order of size as the dimensions of the room. For example, at 100 Hz, the wavelength is 3·35 metres. This means that when sound at these low frequencies is generated in a room, there will be big differences in pressure levels from one part of a room to another, owing to the standing wave patterns that are set up. It is then difficult to decide what the average pressure levels in the room are. Also the partition may itself be about the same size as the wavelength (except in thickness) which makes its sound insulation behaviour erratic. But, these low frequencies are often important subjectively. A reasonable compromise is 100 Hz—this is the lowest one-third-octave frequency band at which sound insulation measurements are usually made. At the other end of the scale, there is little point in going to the highest audible frequencies. It is seldom that these frequencies are important to insulation calculations. Again, a reasonable compromise for the upper limit is 3150 Hz.

Therefore, the 16 one-third-octaves ranged between the centres 100–3150 Hz have been used as a standard test range, i.e. one-third-octaves centred on 100, 125, 160, 200, 250, 315, 400, 500, 630, 800, 1000, 1250, 1600, 2000, 2500 and 3150 Hz. Sometimes octave band figures rather than one-third-octave figures are enough, provided that primary resonances and coincidence effects are allowed for.

Laboratory measurements of sound insulation and site measurements between two reasonably reverberant rooms (e.g. rooms in dwellings, small offices, classrooms) are made as follows. Warble tones or broad-band noise are fed to loudspeakers in one room (the source room). The aim is to make the sound field in this room as diffuse as possible. More than one loudspeaker is therefore suggested. The sound pressure is measured at several positions in the room. As many as ten positions may be appropriate at low frequencies, reducing to six or five at higher frequencies, where pressure variation is less. The readings from the several positions are combined to get the average pressure level (L_1) in the room. The average pressure level should, strictly, be obtained from the arithmetic mean of intensities at each position at each frequency. The arithmetic mean of the decibel values is less accurate, but often adequate, since, in this case, the errors introduced will apply to both source and receiving rooms and will therefore tend to cancel out.

After the pressures have been measured in the source room, the microphones are moved to the room on the other side of the partition being measured (the receiving room). The same noise output is repeated in the source room. The sound pressures are measured in the receiving room as before and the average pressure obtained (L_2). Alternatively, measurements can be made in both source and receiving rooms simultaneously if enough equipment is available.

The level L_2 in the receiving room will depend not only on the Sound Reduction Index (SRI) of the partition, but also on the area of partition common to both rooms and on the amount of absorption in the receiving room. In laboratory tests, precautions can be taken to make flanking transmission negligible and the SRI is then given by:

$$\text{SRI} = L_1 - L_2 + 10 \log S - 10 \log A \ \text{dB}$$

where S is the common area of partition in m² and A is the absorption in m² units in the receiving room. We have already seen that the quantity A can be derived from the room volume and reverberation time, using the Sabine formula. A more convenient form of the above equation may therefore be:

$$\text{SRI} = L_1 - L_2 + 10 \log \frac{S \times RT}{0.16 \times V} \text{dB}$$

where RT is the reverberation time in seconds and V is the room volume (m³).

Example: Suppose the sound pressure level difference across a partition between two rooms is 33 dB at 250 Hz. Then, if the reverberation time of the receiving room is 0·8 seconds at 250 Hz, the area of the partition common to both rooms is 10 m² and the

H*   233

receiving room volume is 100 m³, SRI in the absence of flanking would be:

$$SRI = 33 + 10 \log \frac{10 \times 0 \cdot 8}{0 \cdot 16 \times 100} \text{ dB}$$

$$= 33 + 10 \log \tfrac{1}{2}$$

$$= 30 \text{ dB}.$$

There are two minor differences between laboratory and field measurements of SRI. The size of the partition on site may be substantially different from that used for laboratory testing and the edge fixing conditions may also be different, resulting in some effect on insulation performance. But the major difference between laboratory and site measurement is the important contribution of flanking transmission. For example, a floating floor construction measured in a laboratory may have insulation of the order of 65 dB nominal (i.e. mean SRI 100–3150 Hz). But if the floor is used in practice on 115 mm supporting brick walls (perhaps the inner leaf of a 280 mm cavity wall), then the insulation is likely to be reduced to about 45 dB nominal.

Of course, it is the sound pressure level difference that is of ultimate importance. Existing constructions may be checked for a simple level difference $(L_1 - L_2)$. But often tests are carried out in conditions where changes in the receiving room are to be expected later e.g. in buildings not yet furnished. Two common ways of adjusting the level difference are as follows:

First, a reference reverberation time (for the final receiving room condition) can be used. The corrected value of insulation is then given by:

$$\text{Eventual difference } D = L_1 - L_2 + 10 \log RT - 10 \log RT_{ref} \text{ dB}$$

where L_1 and L_2 are as before, RT is the measured reverberation time of the receiving room and RT_{ref} the reference reverberation time.

Alternatively, a reference absorption for the receiving room (usually 10 m²) can be used. The corrected value of insulation is then:

$$D = L_1 - L_2 + 10 \log 10 - 10 \log A \text{ dB}$$

where A is the measured absorption (m²) in the receiving room.

In housing, it has been found that the mid-frequency reverberation time of furnished living rooms is usually close to 0·5 seconds independent of their volume. This means that if the measured values are corrected to a standard reverberation time of 0·5 seconds, the corrected measured values will be close to the values existing in practice when the rooms are furnished and occupied. The normalised level difference is then given by:

$$D_n = L_1 - L_2 + 10 \log RT - 10 \log 0.5 \text{ dB}$$

where RT is the measured reverberation time during tests. For example, suppose the sound insulation of a floor is to be measured, it would be done over the full range (100 — 3150 Hz), but we will consider one frequency only, say 400 Hz. The test rooms are unfurnished. The sound pressure level readings are as follows:

								Arithmetic mean	
Source room	99	100	95	95	97	98	$L_1 =$	97	dB.
Receiving room	60	55	59	59	63	59	$L_2 =$	59	dB

The measured reverberation time in the receiving room is 1·5 seconds. Therefore the normalised level difference is:

$$D_n = 97 - 59 + 10 \log 1.5 - 10 \log 0.5 \text{ dB}$$
$$= 38 + 5 \text{ dB}$$
$$= 43 \text{ dB}.$$

Note that if the more accurate average of the levels (with reference to intensity p. 289) had been calculated, L_1 would have been 98 dB and L_2 would have been 60 dB, leaving D_n unchanged.

Accurate predictions of level differences cannot be made, if the indirect transmission is not negligible. Instead, if some calculation must be made, we can only use field measurements of the same sort of construction, under the same sort of flanking conditions, to get some estimate of the SRI including the indirect transmission. In this situation, the use of octave bands is often adequate. Appendix C gives typical SRI values for several common types of construction, allowing for some indirect transmission and assuming reasonably good building practice.

Non-Uniform Partitions

If the partition between two rooms is made up of more than one type of construction (each being exposed to the same sound pressure level), then the composite SRI is obtained from the average sound transmission coefficient as follows:

$$\mathscr{T}_{ave} = \frac{\mathscr{T}_1 S_1 + \mathscr{T}_2 S_2 + \mathscr{T}_3 S_3 + \dots\dots\dots}{S}$$

where $\mathscr{T}_1, \mathscr{T}_2, \mathscr{T}_3 \dots$ are the coefficients for each element; $S_1, S_2, S_3,$ ... are their corresponding areas; and ΣS is the sum of all the areas.

The average sound reduction is then $10 \log \dfrac{1}{\mathscr{T}_{ave}}$.

For example, suppose a partition between two rooms is made up of a 7·5 m² area of a material with Sound Reduction Index at a certain frequency of 30 dB and a 2·5 m² area of another material with Sound Reduction Index of 10 dB. Then

$$30 = 10 \log \frac{1}{\mathscr{T}_1} \text{ and } 10 = 10 \log \frac{1}{\mathscr{T}_2}$$

So, $\mathscr{T}_1 = 0\cdot001$, $\mathscr{T}_2 = 0\cdot1$ $\mathscr{T}_1 S_1 = 0\cdot0075$ and $\mathscr{T}_2 S_2 = 0\cdot25$

$$\mathscr{T}_{ave} = \frac{0\cdot0075 + 0\cdot25}{10} = \frac{0\cdot2575}{10} = 0\cdot02575$$

$$\text{SRI}_{ave} = 10 \log \frac{1}{0\cdot02575} = 16 \text{ dB.}$$

A graph for more rapid calculation of the composite insulation of a partition made up of two areas is shown in Fig. 92. The ratio of the area of the element of lower insulation to that of the material of higher insulation (in our example 1:3) is read from the left-hand side and we move across to the right until we meet the curve giving the difference in SRI between the two areas (in the example, 20 dB). We then drop vertically down to the lower scale, to read off the loss of insulation, to be deducted from the higher of the two original sound reduction indices, i.e. 14 dB to be subtracted from 30 dB, leaving 16 dB.

We have, so far, been considering only the insulation between reasonably reverberant rooms. But, in 'dead' receiving rooms, where we cannot expect an even sound field, e.g. open-plan areas, the value of L_2 will drop substantially with distance from the partition. For positions close to the partition, the level difference is likely to be approximately:

$$L_1 - L_2 = \text{SRI} + 3 \text{ dB.}$$

But the change with distance is subject to the gradient of the sound field across the room, which will vary for different room geometry

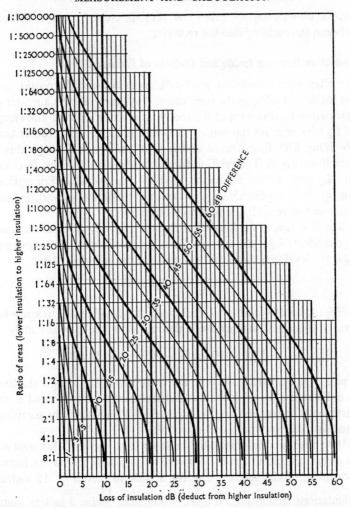

Fig. 92. Performance of composite partitions

and absorption patterns. Usually, we are more concerned to deal with the nearest listening position, often near the partition. To deal with more complicated cases would become too involved to be appropriate here.

It is important to note that where calculations of composite insulation involve very large constructions, e.g. a factory wall with a roller-shutter door opened at one end, it is found to be more accurate to carry out separate calculations for the component parts, since the

237

overall dimensions are often not very different from the distance between the building and the receiver.

Insulation Between Inside and Outside of Buildings

We often need to estimate level differences between the outside and the inside of buildings. In such cases, it is nearly always a matter of calculating L_2 (the level at the receiving position) from a knowledge of L_1 (the level on the source side) and a known SRI, rather than obtaining SRI from measurement of L_1 and L_2 (although this is sometimes done). The relationships given below show how to calculate L_2 and cannot always be used to measure SRI. For various reasons it is impossible to give accurate relationships and, in each case, we will provide the best approximation.

The first case is the insulation from a relatively reverberant room to outside. If L_1 is the average sound pressure level in the room and L_2 is the level very close to the outside of the partition, then:

$$L_2 = L_1 - SRI - 6\,dB$$

For indoors to outdoors calculations, we usually want to know the level at a certain distance from the partition, i.e. approximately

$$L_2 = L_1 - SRI + 10\log S - 20\log r - 14\,dB$$

where L_1 is the average level in the reverberant room, L_2 is the level at a distance r (in m) from the partition along a line normal to the partition (r is assumed large compared with the size of the partition) and S is the area of the partition in m^2.

If we wish to assess the radiation from large surfaces, this must not be attempted closer than a distance of $D/3$ where D is the largest dimension of the surface; e.g. for a wall 10 metres $\times$ 15 metres, calculations should not be made closer than $\dfrac{15}{3}$ i.e. 5 metres, since, close to the source, the receiving position does not receive significant contributions from the more extreme parts of the source, and complex cancellation patterns in the 'near field' are likely. Under these circumstances, the inverse square law is not applicable.

In practice, it is very often 'weak links', such as a window in a brick wall, that determine the level difference. For such small sources, the relationship above can then be directly applied. Conditions are a little more complicated if there is more than one path for the sound to travel from inside to outside, e.g. via a window and a door, but it is only necessary to calculate two separate values of L_2 from the two values of SRI and S and then add these two values (as described on p. 212).

If the receiving position is not on the normal to the radiating area, a very approximate estimate of the reduction due to the departure from the normal can be found in Table X for two areas of opening and angles of 45° and 90° from the normal.

Table X Reduction dB at side position compared with normal position.

| Octave band centre frequency (Hz) | Area of opening | | | |
| | 1 m² | | 10 m² | |
	45°	90°	45°	90°
63	0	1	1	2
125	0	3	3	5·5
250	1·5	5·5	6	9·5
500	3	8	8·5	11
1000	5	9	9·5	13
2000	6·5	10·5	11	15
4000	8	12	13	18
8000	8·5	13	15	19

For calculations concerning noise travelling outside to inside, the following relationship may be used if the room involved is relatively reverberant:

$$L_2 = L_1 - \text{SRI} + 10 \log S - 10 \log A + 6 \text{ dB}$$

where L_2 is the reverberant sound level in the room, L_1 is the level at one metre from the façade of the building, S is the surface area (m²) of the perimeter construction, A is the total absorption in the room (m²).

We can make an approximate estimate of the sound pressures on the surfaces of a building not directly facing the noise source, assuming that the distance to the noise source is large compared with the building dimensions. For example, from Fig. 93 Façade A can be treated on the basis of the above equation. Façades B and C are likely to receive levels approximately 6 dBA below those on Façade A. Façade D is most difficult to assess because reflections from other buildings are likely to influence this situation strongly. However, a

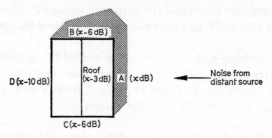

Fig. 93. Typical variation of exposure to noise of
façades of a building

level 10 dB below the exposure of Façade A is likely. A pitched roof
offers a reduced projected area to the source and a reduction of
3 to 4 dB in exposure is a likely result.

If the sound is coming vertically, i.e. from aircraft in flight, then
for a flat roof the above equation is appropriate; for a pitched roof
2 to 3 dB less is likely and for the walls again about 2 to 3 dB less.

Enclosures

Estimates of noise reduction by enclosures can only be very approximate. Difficulties arise in assessing the complex sound field within
the enclosure, the adjustments to source directivity, the behaviour
of any absorbent material lining the inside of the enclosure and the
sound insulation performance of materials used in this form.

It would not be helpful to the reader to suggest accurate calculation procedures for enclosure performance, since, in practice, it is
more useful to refer to field data in the form of sound pressure level
reduction achieved by specific forms of enclosure geometry and
construction. But two particular warnings must be given. First, an
enclosure used close to a wall or roof will 'couple' with it, so that the
overall insulation will not be the sum of the reductions which might
be expected for each barrier separately. Second, it is not sufficient
to calculate L_2 on the basis of a reduction equal to the SRI figures
alone. An enclosure collects sound energy inside, resulting in a local
build-up of noise around the source (L_1). Estimates of L_2 must
therefore allow for the build up in L_1. This build-up can be countered
by using absorbent linings inside enclosures, as is usually done. But
the treatment normally provides little absorption at low frequencies.
For typical lightweight mechanical plant enclosures (e.g. sheet metal
with 25 mm or 50 mm absorbent lining inside) the reduction in noise
is usually less than might be expected from SRI figures at low fre-

quencies, often by about 5 dB. At high frequencies, SRI figures often underestimate the performance.

Measurements of enclosure performance will normally be carried out on a substitution method (i.e. with and without the enclosure), in an acoustic laboratory, or preferably at the position of interest. Measurement of L_1 and L_2 on site will not yield a direct performance figure for the enclosure, and manufacturer's data presented on enclosure performance should be checked against this possibility.

Impact-Sound Isolation

By impact-sound isolation, we mean isolation against noise generated by impacts on the structure e.g. footsteps. (Vibration, as defined on p. 150, is not included here.) Most impacts occur on floors and we will confine ourselves to this case.

Impact-sound isolation is measured using a standard 'impact' or 'footsteps' machine. This produces impacts of standard energy on the surface of the floor being measured and the resulting sound pressures in the room below (the receiving room) are measured. It should be stressed that, unlike airborne sound insulation measurements, which are relative, an absolute measure of sound pressure is now required, needing a calibrated microphone.

The pressure level measurements in the receiving room are made at several positions and the results averaged. At least two positions of the source should also be used. As with airborne sound insulation, the average of the intensities should be taken. Results are usually plotted in one-third-octaves.

Laboratory and field measurements usually differ sufficiently for the former to be of limited use for prediction of absolute impact-noise levels. This is due to variation in flanking transmission, floor size and/or edge details. But the laboratory can be particularly useful in assessing the improvement in impact isolation (e.g. as a result of adding a resilient covering to a floor construction). This is then a comparative (before and after) test yielding a set of data referring to improvement in impact isolation.

Measured values of impact noise are usually corrected or normalised with reference to standard absorption of 10 m² or a reference reverberation time, as referred to on p. 234. For example, normalisation against a reference reverberation time will be:

$$L_n = L + 10 \log RT_{ref} - 10 \log RT \quad dB.$$

If compared with the equation on p. 234, it will be seen that the absolute nature of impact measurement is expressed here as L compared with the $L_1 - L_2$ in the equation on p. 234. With the same adjustment, the other methods of normalisation referred to on pp. 234–235 may equally be used for impact-noise measurement.

Calculation of impact noise is complex and rarely appropriate, and will not be dealt with here.

Mechanical Services

Calculation and measurement of noise from mechanical services systems have become a great deal more complex as the practice of noise control has developed, and the variety of equipment and systems has expanded. For detailed methods of calculation and measurement, the reader is advised to refer to textbooks, which cover this field in more depth than space allows here. Although we will not give comprehensive means of calculation, some of the more important relationships are given as an extension of the principles given in Chapter 6, with an emphasis on mechanical ventilation, which needs a slightly different approach.

Mechanical Ventilation Systems

So far in this book it has been appropriate to refer to sound pressure level as the measure of sound most appropriate to our needs. But in dealing with mechanical plant, it is often more convenient to describe the noise output of a source, independent of its location. The sound pressure level produced by a noise source will vary, depending on whether it is placed in, e.g. a small, well-insulated box where the pressure will build up, or in the open air. Therefore the reader may well come across references to sound power output rather than sound pressure from an item of plant. We will not become involved with the detailed use of sound power, but it is important that the reader understands the difference between power and pressure, so as to be able to distinguish between these when sound data is presented. As a parallel to the case of the sound in a small, insulated room or radiated into the open air, we can consider the temperature arising from the use of a heat source varying, depending on whether we are in a small, thermally insulated room, or out of doors. In these cases, although the sound pressure (or the temperature) is usually the quantity we need to calculate eventually, the quantity which describes the noise (or heat) output of the source is its power. Therefore we will come across references to the sound power of a source. For convenience in relating sound power to sound pressure, the decibel is also used to describe power. This infers a relative level, as we have seen in Chapter 1. In the case of sound power, the reference standard is 10^{-12} watts and the derivation of the units is such that a change in sound power produces a corresponding change in sound pressure in decibel terms. Data related to mechanical sources may be presented as sound power (total) or sound power (in duct) or sound pressure level at a given distance. All are presented in terms of decibels. There is therefore a danger of considerable confusion and error, unless the quantity is carefully specified with its reference value. Occasionally, noise data for mechanical plant is presented as a sound pressure level, without even any indication of distance from the source or the room (if any) conditions. In this case, more information is clearly needed.

Fan Noise

The over-all noise output of a fan is usually related closely to its duty, i.e. its air volume flow and the air pressure drop across it:

$$\text{Noise output} \simeq 10 \log Q + 20 \log p + K \ \ dB$$

where Q is the volume flow and p is the pressure drop across the fan. In practice this means that a doubling of pressure results in an increase of 6 dB, a doubling of volume results in an increase of 3 dB. But the spectrum of noise produced will vary for different types of fan. Fig. 94 shows the typical spectrum shapes applicable to axial and centrifugal fans. It will be seen that at the lower frequencies (usually more difficult to attenuate) centrifugal fans tend to exhibit the higher noise output. But it is generally unrealistic to select fans

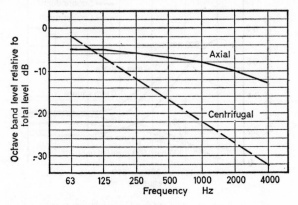

Fig. 94. Typical fan noise spectra

243

solely by their noise spectrum. There are usually greater priorities which determine the form of fan which is used. Experience shows that a fan of a given duty is quietest if it has been selected for a duty in the range where it is operating most efficiently. If this is not the case, for example, if a fan is forced into a part-stall condition by the load presented to it, noise levels will rise sharply. The inlet to the fan is also important. If, as a result of tortuous or abrupt air-intake conditions air arrives at the fan in a turbulent state, noise levels are likely to rise by up to 10 dB.

In some cases, e.g. with paddle-blade fans, the running speed may result in a strong tone—a drone or whine, which must be tackled separately from the broad-band noise.

Duct-borne Noise

The attenuation of sound in ducts depends on their shape and size, whether or not they are lined or whether they have purpose-designed attenuators fitted. Circular ducts, which are relatively rigid and do not move easily in response to excitation by sound provide little attenuation with distance, i.e. the noise is kept inside the duct. But rectangular ducts will move more easily and tend to allow noise to escape along their length, resulting in higher attenuation as seen within the duct. Fig. 95 gives some typical values for attenuation by straight, unlined ducts. The effects of lining the ducts with acoustically-absorbent material are greater for small ducts (see Fig. 96). Where ducts are large, and even when they are not, the use of absor-

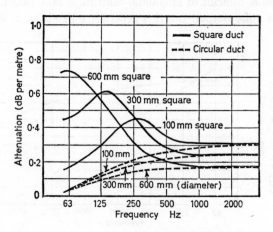

Fig. 95. Typical attenuation along unlined ducts

244

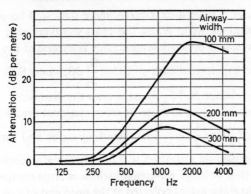

Fig. 96. Typical attenuation due to lining ducts with
e.g. 25 mm mineral wool

bent material set in the airway (e.g. a splitter or pod attenuator) may
be more economical.

A bend in the duct will result in attenuation, which increases if
the bend is lined (Fig. 97). As the air is distributed by passing into
the various branches of the system, the noise can generally be taken
to split in the same proportion as the air. For example, if, at a branch,
two-thirds of the air travels in one direction d_1 and one-third in
direction d_2, the attenuation at the branch would be:

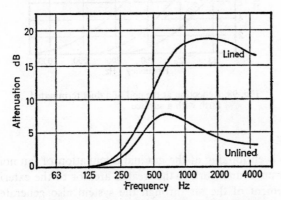

Fig. 97. Typical attenuation at a right-angled bend in a
300 mm square duct, unlined, and with a 25 mm
mineral-wool lining

	Attenuation
along path d_1, taking $\frac{2}{3}$ of the air	2 dB
along path d_2, taking $\frac{1}{3}$ of the air	5 dB.

However, at a branch or T-piece, there may well be attenuation due to a bend as well. When the cross-sectional area of a duct changes, the sound pressure level change is expressed as:

Sound pressure level change $= 10 \log \dfrac{S_1}{S_2}$ dB, where S_1 is the area before the change and S_2 the area after the change.

An expansion of a duct to twice its cross-sectional area would result in a change of $10 \log \dfrac{1}{2}$ i.e. $-$ 3dB. Standing waves and turbulence in expansion or plenum chambers will modify the attenuation. But, in addition to this, noise of wavelength which is large compared to the duct size is to some extent reflected at the open end of a duct. So, the smaller the duct, the more noticeable is this effect. Fig. 98 gives an indication of the order of attenuation which occurs in this way.

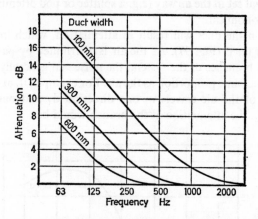

Fig. 98. Reflection at the end of a duct terminated
flush with a surface

Air Noise

We have been looking at the gradual attenuation of fan noise as it passes through ductwork to the treated area or to the exterior. But the movement of the air through the system also generates noise (particularly in high-velocity systems), which may pass through to the end of the system. The noise is created by the interaction between turbulent air and, in particular, solid bodies in the air flow, e.g. dampers, grilles and diffusers.

$$\text{Noise output} \simeq 10 \log S + 60 \log V + 30 \log p + K \ \ dB$$

246

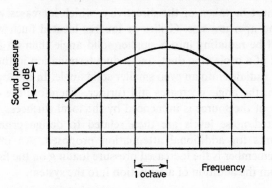

Fig. 99. Typical spectrum due to air noise in ventilation
systems

where S is the surface area of the obstruction, V is the approach velocity and p is the pressure drop across the obstruction.

Typically, the spectrum shape of air noise is as shown in Fig. 99. The position of the peak in the spectrum is related to a representative dimension of the obstruction, i.e. the larger the obstruction, the lower the peak frequency.

Very often suppliers of mixing boxes, grilles and diffusers will provide noise levels for their equipment, based on laboratory or field tests. Usually these tests are carried out with even air flow on to the units, and care must be taken to allow for more noise if less even conditions are likely, e.g. turbulence and/or noise generation from convoluted flexible duct connections. Data for grilles, diffusers and induction units is often given in terms of sound pressure levels, NR or NC criteria, after assuming a stated quantity of acoustic absorption in the room to be served. This data usually refers to the reverberant sound level and care must be taken to see that direct sound levels are not excessive. Corrections may also be necessary to ensure that reverberant levels are based on the actual and not the assumed absorption. Since many of these items are used in large numbers, the contributions from individual units must be added to obtain the over-all noise level.

It is worth pointing out that such noise generated right at the end of the system cannot be attenuated. It must be removed as part of the design and commissioning of the system.

Once the noise reaches the room or the atmosphere, the terminal opening can be considered a new source of noise and direct and reverberant (if applicable) contributions can then be derived. Some account must be taken of the directivity, which increases with frequency and aperture size. Grilles or louvres located flush with e.g. a wall will be radiating into a smaller solid angle than an aperture at the end of a projecting duct, and a duct termination in the corner of a room radiates into an even smaller solid angle. In the top corner of a room, the angle decreases still further. Therefore the effective directivity of the source is influenced by the local surfaces.

Calculated noise levels are then related to design criteria and requirements for additional attenuation proposed. An important point to remember is the increased pressure loading on the fan which results from the addition of attenuation into the system.

Measurements in Ducts

Noise measurements inside ducts are complicated by the air movement and the restricted space. A wind shield must be fitted and securely attached to the microphone. Even so, there will be a limit to the air velocity in which this shield is effective. Often gaining access to the duct, air handling unit or plenum chamber, means changing the loading on the fan, thereby altering the noise generation, unless the access can be closed during the measurement. The accuracy of measurements, particularly at low frequencies, is limited owing to the small spaces involved. In practice, the most successful tests on noise in ventilation systems are achieved by substitution methods, concentrating on measurement of noise which is fed into the room or laboratory.

Room-to-Room Transmission via Ductwork

Ductwork systems which are common to a number of rooms can provide paths for other noise between the rooms (see Fig. 75, p. 197). The reverberant sound pressure level in the receiving room is given approximately by:

$$L_2 = L_1 - att_{duct} + 10 \log S - 10 \log A \quad dB$$

where L_1 is the reverberant sound pressure level in the source room, att_{duct} is the attenuation through the duct, S is the cross sectional area of the duct (m^2) and A is the total absorption (sabins in m^2) in the receiving room. Attenuation through the duct will include attenuation due to reflection at the end of the duct (see Fig. 98). Also, if the listener is close to the duct termination, direct sound will produce higher levels locally.

248

Noise Transmission from Plant Rooms

Plant rooms are often the noisiest areas in buildings and calculation of noise transmission to adjacent rooms (see p. 233) is required to be as accurate as possible to avoid noise problems, but not to over-design.

Unfortunately, the difference in sound insulation between, say, 150 mm and 300 mm reinforced concrete floors is only about 5 dB, although the effect on the structure, particularly high up in the building is considerable. But it must be clear that estimates of noise output of equipment (which often have to be made before it is finally selected) are rarely accurate to less than $\pm$ 5 dB, and a number of other errors can creep into the calculation, so that the confidence limits are wide and the value of detailed calculation is somewhat limited.

Even so, experience of plant-room noise has grown and calculation in combination with experience can provide reasonable guidance. Currently, much of the information about noise sources is given in terms of sound power (see p. 234) e.g. for fans, and corrections have to be made for the losses through the sheet-metal casings, introducing more steps in the calculation and potentially more error. But more accurate calculations are possible based on source data in terms of a sound pressure level spectrum at 1 metre, such as is often presented for boilers or compressors. The critical reverberant sound pressure level needed to estimate insulation is very often close to the levels at 1 metre distance anyway, if the distribution of the plant within the room is of average density. But, as yet, more suppliers of equipment have to be persuaded to present data related to noise radiation into the plant-room as well as in ducts.

For the most part, the direct sound from plant does not influence insulation calculations very much, since the surface area affected is usually small, where direct sound levels are particularly high. But, occasionally, where very large, noisy plants are positioned directly over or alongside a small, relatively 'live' room, the direct radiation through the floor (or wall) may become important. The build-up of noise inside air-handling units can also lead to this situation. But usually, the floor slab is thickened locally to form a plinth/house-keeping pad which will assist insulation. Sometimes, the base of an item of plant must be floated on a concrete inertia base to improve insulation and vibration isolation.

When measuring insulation between plant rooms and adjacent

areas, a check should be made on likely flanking paths, e.g. via ducts, service shafts, doors or pipe apertures. Insulation is very often limited by these paths rather than the separating structure itself.

Assessment of noise leaving plant rooms via e.g. fresh-air louvres can be carried out using the method described on p. 238.

Pipe-borne Noise

Attenuation of sound along pipework rarely exceeds 0·5 dB per metre and is more typically in the range 0·05 to 0·1 dB per metre. Natural attenuation along pipework is therefore not generally subjected to detailed calculation. Because of the long distances of propagation, reduction of the amount of energy being fed into the pipes or keeping the pipe-borne energy out of the structure are most important (see p. 148).

Vibration

Owing to the increasing complexity of practical vibration calculations and measurements, and the empirical nature of much of the vibration design data, we will not attempt to cover more than the principles described in Chapter 6.

250

10

Noise Control in Practice

In this chapter, we hope to provide a framework on which the reader can build his or her own practical experience. We will describe and comment on some of the more common techniques used in practice by referring to various elements in buildings which play a prominent part in noise control.

Noise control measures must fit in with good building practice and we can expect that conflict will sometimes arise between the needs of noise control and constraints such as cost or delivery dates for materials. However, it is important not to compromise too far. Although 'a few dB' tolerance may be acceptable on occasions, this approach if repeated will clearly lead to poor results. Methods must also be feasible with the resources available and for best results should be planned in good time, carefully detailed and supervised as described in Chapter 8.

Windows

(a) Opening Windows

If windows are likely to be opened often, there is little point in spending a lot of money on sound-insulating glass, that is, unless we want to give the occupant the option of shutting out external noise for short periods. Many existing buildings suffer noise because windows have to be opened to relieve solar heat gain—the occupant has the choice of the heat or the noise. Even so, a decision to seal glazing and introduce mechanical ventilation is an expensive one and cannot be taken lightly. But an open window gives little protection—5 to 10 dB is typical. Some windows need to open only occasionally e.g. for cleaning, or as smoke vents. Normally, these can be closed on to good seals and can be kept locked for most of the time.

Many windows which are intended to be closed are partly open or poorly sealed when closed. The concern to save heat has helped to improve seals on opening windows. But ill-fitting frames (e.g. metal frames painted many times, or sliding sashes with wide tolerances) are still common. In general, there is a better chance of airtightness (and therefore improved sound insulation) with compression seals rather than, for example, brush seals. But to be effective, ironmongery must be capable of forcing the compression necessary to make the seal. Casement windows (which normally use compression seals, if any) can therefore be sealed fairly easily. Pivot windows are awkward, because the line of the seal must be maintained across the pivot, or two separate lines of wiping seals are used (see Fig. 100).

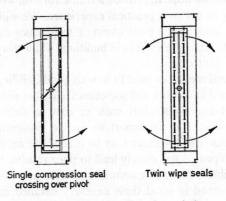

Single compression seal
crossing over pivot

Twin wipe seals

Fig. 100. Two ways of sealing pivot windows

This is only successful with careful detailing and manufacture to close tolerances. Sliding windows usually move along their seals, making it very difficult to keep airtightness, without an elaborate and separate sealing detail.

Single windows (say 4 mm or 6 mm glass) which are not well sealed tend to give insulation in the range 15–20 dB.

(b) Sealed Single Glazing

If we assume for the moment that windows are well sealed and carefully detailed, we can turn our attention to the capabilities of the glass itself. Because of its high densities, even thin glass can provide good sound insulation. Well sealed single glass gives insulation ranging from about 23 dB, for 3 mm glass, up to 33/34 dB for 25 mm glass. Performance above 35 dB can be reached with very thick (> 12 mm) laminated glasses (see Table XI). There is some evidence that mounting glass in resilient material, e.g. neoprene, offers a small advantage over the use of a more rigid fixing, e.g. putty.

Coincidence losses (see p. 139) are particularly important for single

252

Table XI Average airborne sound insulation

This table indicates the airborne sound insulation, averaged over the frequency range 100 to 3150 Hz, of a number of common types of constructions. It is assumed that there are no holes or cracks in the constructions, except those specifically mentioned.

As has been pointed out in Chapter 6, these single-figure values must be taken only as a guide because insulation effectiveness depends on how the insulation varies with frequency and because differences in building construction affect the values actually obtained.

It must also be remembered that the insulation achieved in practice depends not only on the insulation of the particular dividing element but also on its area in relation to the sound absorption in the rooms, and on indirect transmission. No specific allowance can be made for indirect transmission, except to say that for elements having an insulation of 40 dB or below it will have little effect. In the following tables the figures above 40 dB allow for the amount of indirect transmission likely to be present when the structures are used in a more or less traditional manner.

As to the effects of area and absorption, the values given have been chosen to represent as nearly as possible the achieved insulation between two normally furnished rooms of average proportions when the whole area of the wall or floor is of the specified construction.

As a result of these considerations, and owing to variations in detailing and quality of workmanship, it is important to bear in mind that the figures in these tables are subject to wide tolerances and must only be treated as estimates. Estimates for other solid, sealed wall or floor constructions which do not appear in this table can be made from the 'Mass Law' curve, Fig. 52. A table of the weights of various common building materials is given in Appendix B.

Windows

Open window	5–10 dB
3 or 4 mm glass, not well sealed	15–20
3 or 4 mm glass, well sealed	23
6 mm glass, well sealed 4 mm glass—200 mm gap—4 mm glass, but each leaf partially opened (up to 100 mm opening), with openings staggered 1·5 metres apart	27
6 mm glass—12 mm gap—6 mm glass well sealed	28
12 mm glass, well sealed	31
6 mm glass—150 mm gap—4 mm glass well sealed, lined reveals	35
12 mm glass laminated, well sealed 4 mm glass—200 mm gap—4 mm glass, but not well sealed	36
10 mm glass—80 mm gap—6 mm glass well sealed, lined reveals	37
4 mm glass—200 mm gap—4 mm glass well sealed, lined reveals	39
10 mm glass—200 mm gap—6 mm glass well sealed, lined reveals	44

Masonry walls	
Lightweight blockwork, not sealed	<35B
63 mm hollow clay block plastered on each side to 12 mm	35
57 mm lightweight (110 kg/m²) blockwork, plastered both sides to 12 mm	39
50 mm precast concrete units, well grouted joints 200 mm lightweight concrete precast slabs (122 kg/m²) with well grouted joints	40
100 mm solid brickwork, unplastered	42
—plastered 110 mm dense concrete, well sealed	45
150 mm dense concrete, well sealed	47
230 mm solid brick, unplastered	48
250 mm no-fines concrete (1:10) plastered both sides to 12 mm Blockwork (110 kg/m²)—75 mm air space—blockwork (110 kg/m²)	49
200 mm dense concrete blocks with cement-lime mortar plastered both sides to 12 mm 230 mm solid brickwork plastered or with dry lining of 12 mm plasterboard on plaster dabs 250 mm cavity construction, i.e. 110 mm brick—cavity 75 mm block + 12 mm plaster, butterfly wire ties	50
280 mm cavity brickwork, butterfly wire ties, plastered one side to 12 mm	52
340 mm brickwork, plastered both sides to 12 mm	53
450 mm brick or stone, well pointed or plastered	55

These estimates refer to insulation of walls between rooms. For insulation between a room and a noise in the open air the overall insulation should be *reduced* by 5 dB.

Floors	
21 mm t- and-g boards or 19 mm chipboard on floor joists, 9 mm plasterboard + skim coat below	35 dB
110 mm concrete + screed (⊁ 220 kg/m²)	42
21 mm t- and-g boards or 19 mm chipboard on floor joists, lath and plaster (20 mm) below with 50 mm layer of sand pugging	45–46
125 mm reinforced concrete + 50 mm screed	
200 mm reinforced concrete + 50 mm screed 125 mm reinforced concrete + timber raft (i.e. 21 mm t-and-g boards or 19 mm chipboard) on glass- or mineral-fibre quilt Precast concrete units (50 mm) + 30 mm finishing screed on deep truss with well sealed 9 mm insulation board below	47–48
125 mm reinforced concrete + 40–50 mm concrete screed on glass- or mineral-fibre quilt 300 mm reinforced concrete + 50 mm screed Floated timber raft (21 mm t-and-g boards or 19 mm chipboard) on glass- or mineral-fibre quilt on joists + 50 mm sand directly on ceiling of plaster on metal lath (20 mm)	49–50
150 mm reinforced concrete + 100 mm floated raft on specialist mounts, free from 'bridging'	55
As above, with walls built near the edge of the floor to limit flanking transmission	60+

Dry partitions

21 mm t-and-g boarding, tightly clamped	20 dB
9 mm insulation board each side of 75 × 50 mm studs	23
12 mm plasterboard + skim coat on timber frame	25
12 mm plasterboard both sides of honeycomb core—63 mm total thickness	28
6 mm ply/hardboard on 50 mm timber studs, mineral-fibre quilt in cavity 12 mm plasterboard each side of 50 mm timber studs 1 mm steel panels spaced apart by 50 mm with mineral-fibre quilt fill	30
12 mm plasterboard each side of 50 mm metal studs	33
50 mm wood-wool plastered both sides to 12 mm	35
12 mm plasterboard each side of 50 mm timber studs with absorbent quilt in cavity	37
As above, but with metal studs	39
2 × 12 mm plasterboard each side of 50 mm timber studs with absorbent quilt in cavity 1 mm steel panels backed with 9 mm plasterboard, with absorbent quilt in cavity	41
2 × 12 mm plasterboard each side of 75 mm metal studs with absorbent quilt in cavity	45
3 layers 12 mm plasterboard on timber frame each side of 225 mm air gap, frames separated and area of supported panels ≯ 15% glass- or mineral-fibre quilt hung in cavity	49

Doors

Panel/hollow-core door—well fitted, no seals	15 dB
—as above with good seals and close-cut threshold	20
Solid-core door—well fitted, no seals	15
—as above with good seals and threshold strip seal or close-cut to carpet	25
60 mm + solid-core door—with carefully detailed seals, including threshold seal	30

Above 30 dB, specialist doors are needed.

Room to room insulation via suspended ceiling void

assuming 600 mm deep void, below concrete soffit, no undue obstructions such as large ducts, downstand beams.

>10% perforated metal pan ceiling with absorbent lay-in backing	15–20 dB
19 mm mineral-fibre ceiling tile (6 kg/m²) in lay-in or concealed-fix grid	25–30
Solid metal pan ceiling (0·6 mm) + absorbent quilt overlay	about 30
Perforated metal pan + 9 mm plasterboard backing	about 35 35–40

glass, occurring at the upper end of the audio-frequency range. A check should always be made to see whether this weakness is significant for the case in hand. The thicker the glass, the lower is the coincidence frequency. For 25 mm glass, coincidence can be as low as 500 Hz, if the angle of incidence of the sound is near to the normal.

Laminated glass tends to give higher sound insulation, partly because weakness at the coincidence frequency is less pronounced. This is because of the inherent damping resulting from the multi-layer construction.

(c) Double Glazing with a Narrow Air space

Two thin sheets of glass separated by a narrow gap are not as good for sound insulation as heavy single glass, particularly at low frequencies. This is because the double window is not as stiff. If both panes are of equal thickness, coincidence losses will be superimposed on one another and will show up strongly. Therefore windows of this type of construction often allow a narrow band of noise to pass through, which is heard as a 'hiss' inside. It is good practice, particularly with narrow double glazing, to spread the coincidence frequencies apart and so allow each pane to cover for the coincidence loss of the other, i.e. by using different weights of glass.

Narrow double glazing may be convenient to install and to provide the necessary thermal insulation. Unfortunately, it offers little advantage as far as sound insulation is concerned.

(d) Double Glazing with a Deep Air space

A deep air space is needed (50 mm minimum, often 100 mm or 200 mm) for the sound insulation of double glazing to be significantly better than single glazing. Frequently, there are practical advantages in avoiding a very deep air space. With a gap of 50–80 mm, double glazing gives an improvement of some 3 or 4 dB over the same weight of single glass—this is often enough. If so, the window is still narrow enough to be hung as one unit (even if the panes are set in separated frames linked together across resilient seals). This allows smoke vents and cleaning to be arranged fairly easily.

But, of course, there are many cases where an even deeper airspace is needed i.e. to meet performances between 37 dB and 50 dB, particularly where insulation is needed at low frequencies. As performance improves flanking transmission through frames becomes more relevant. In these cases (for, say, 40 dB+) frames should be well separated. Middle- and high-frequency insulation can be helped by putting absorbent lining in the reveals, e.g. carpet, mineral-fibre board (see Fig. 101). Care must be taken to avoid wedging rigid absorbent board between separate frames, thereby helping flanking. Again, it is useful to vary the thickness of panes to reduce coincidence effects.

Secondary glazing is often installed behind existing windows to improve sound insulation. To make the best of this method, the existing windows should be given good seals first. This may give up to 10 dB improvement immediately. Very often, sliding panes are preferred for the secondary glazing. Although not quite as effective as well sealed casements, they are likely to raise insulation by a further 5–10 dB, if a deep air-space is allowed (> 100 mm). This over-all improvement of 15–20 dB is often adequate.

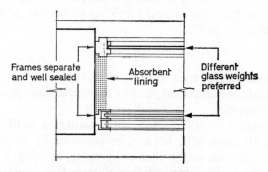

Fig. 101. Lined reveals in double glazing

High insulation can be achieved (up to 30 dB) with open double glazing where the sound is forced to follow a devious path, e.g. a 50 mm slot opening at the bottom of the outer pane and a similar one at the top of the inner pane. This allows some permanent ventilation whilst maintaining reasonable sound insulation.

The overall sound insulation of a construction which includes windows must take account of the ratio of the window area to that of the surrounding construction. Mullion and transom details may not be critical if they occupy a small surface area and are well sealed. But large hollow sections may need packing or filling to improve insulation.

External Walling

Traditional brick, block or concrete external walls tend to give adequate insulation for all but the most severe requirements (unless they contain obvious weaknesses, e.g. open vents or doors). But increasing use of lightweight cladding panels leads to lower insulation values. Particular difficulty can arise with highly resonant materials such as aluminium sheet or glass-reinforced plastics. Panel and cavity resonances associated with them can be excited by individual sources—lorries, buses, trains—as happens occasionally with glazing. This is particularly noticeable if the inner skin is of light panel construction and/or poorly sealed. Damping (see p. 273) should be applied to large panels and the inner skin should be substantial (preferably in brick or block) and well sealed. The use of absorbent material hung in the cavity may be helpful, subject to a check on its effect on thermal insulation and condensation.

The possibility of noise transmission from floor to floor via cladding (or the void behind it) should be checked. Means of taking air through external walls without providing an efficient path for noise are discussed on p. 275.

Roofs

Sound insulation of roofs is most often needed to keep out noise from aircraft or mechanical plant or to contain noise from function rooms or factories.

Traditional domestic pitched-roof construction is likely to give about 35 dB mean SRI, but can be improved by using a sealed, boarded membrane on the pitch or by laying, or fixing, boards over the ceiling joists. Blocking of unused chimney flues may be necessary for older properties.

Basic asbestos-cement or galvanised-metal roofing, well sealed, will give about 25 dB, but normally with thermal insulation and a sheeted soffit (e.g. plastic or foil-faced plasterboard) performances well over 30 dB are possible.

Flat-roof constructions offer a wide range of sound insulation. In estimating performance, care must be taken not to apply the Mass Law to constructions which are porous, e.g. unscreeded wood-wool slabs or non-homogeneous, i.e. of varying surface weight. Many composite constructions must be treated as double-skin partitions and the use of absorbent quilt to damp cavity resonances may be appropriate. It is important to avoid the introduction of condensation as a result of noise control, bearing in mind that many sound-absorbent materials also give thermal insulation, yet are pervious to water vapour. For complex constructions sound insulation should be checked by laboratory tests.

In critical cases, e.g. when protecting studios or auditoria from aircraft noise, it may be necessary to use both mass and separation to reach a suitable performance. This is normally at considerable expense.

For all these cases, obvious weaknesses e.g. roof lights, permanent vents, unattenuated roof extract fans, will need special attention where sound insulation is important.

As with floors (see p. 151), roof construction may affect vibration isolation measures for roof-mounted plant.

Isolation of Buildings

Where very sensitive buildings are to be close to powerful sources of vibration (e.g. an hotel over a railway line), it may be necessary to separate part or all of the building from the ground. This may be expensive, but sometimes the costs compare well with losses which vibration can cause.

To do this, we have to consider the stability of a sprung building. A break at or near ground level means that more than usual consideration must be given to vibration induced by wind and to cross-bracing the structure at low level. Services must be brought in with flexible connections and a wide gap or resilient seal left around the building if it is separated at ground level (see Fig. 102).

The design is based on the level and type of vibration to be isolated, ground conditions and estimates of the vibration response of the

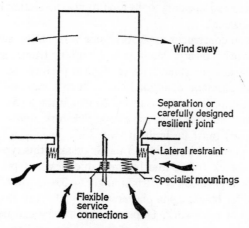

Fig. 102. Diagrammatic section through an isolated building

259

building, and should be carried out with specialist advice. Usually, the aim is to lower the natural resonant frequency of the sprung building to at least one-third of the frequency of vibration. This often results in a need for mounts with a natural frequency in the range 5–8 Hz. Usually, wind excitation can be dealt with by building enough damping into the mounts.

A number of materials have been used for such mounts, with mixed success. Helical steel springs or laminated rubber blocks (rubber layers bonded together with steel plates between them) are perhaps as effective as any, although there is scope for using other new materials. The deflection of steel beams could also form a spring. Design is based on loading, maximum deflection, adequate life and maintenance. Tight specification is needed to overcome potential problems of (where applicable) creep, chemical attack, fire resistance, etc. Details usually allow an arrangement for supporting the building if mounts fail in any way.

Buildings can be built on blocks and then jacked on to mounts. But usually it is easier to build directly off the mounts, keeping a uniform distribution of load. Even if loading is somewhat uneven, the deflection is often no more than might be expected from natural settlement.

Concrete and Masonry Walls

Because of their inherent weight, concrete, stone, bricks and blocks can be used to good advantage for sound insulation. In situ (or precast) concrete and masonry walls tend to give insulation in the range 35–55 dB nominal.

For a given construction, performance varies a lot, subject to the quality of workmanship, detailing and flanking transmission via the associated structure. Table XI gives typical performance figures for a number of common constructions, built to a reasonable standard of workmanship. It will be seen that to match the 50/52 dB of traditional 230 mm solid or 280 mm cavity brickwork, lightweight blocks (60–100 kg/m²) must be built in two skins, usually with a 75 mm cavity and plastered both sides. In a single skin this type of cellular or aerated block (of 75 mm/100 mm thickness), if plastered, is a useful internal partition construction giving 39–43 dB. Slightly more dense or wider block, again plastered, will match the 45 dB of traditional 115 mm brickwork. Unplastered masonry can be very poor (< 25 dB).

For best results, joints in brickwork and blockwork must be full and well pointed. Because of the porosity of some materials and the danger of poor mortar joints, the application of wet plaster, which adds weight and seals the wall, often produces remarkable improvement in sound insulation. For normal (not lightweight) concrete and dense, well-pointed brickwork, plaster is not so good. Where dry linings are to be added, walls should be well sealed behind them. Drying shrinkage may cause cracks which could affect sound insulation. In critical cases, cover strips can be detailed to cope with this. Cast in situ concrete walls must be well compacted and plugged where formwork ties have been used. Precast units need carefully detailed joint seals.

For a given surface weight, cavity construction gives slightly better insulation, particularly at high frequencies. But insulation and flanking transmission via double walls are affected adversely if the two skins are tied together rigidly. If possible, wall ties should be resilient, e.g. butterfly wire ties. Use of cavity fill for improved thermal insulation can reduce sound insulation because of the increased coupling between the skins. But the effect is usually small.

Floors

(a) Suspended Floors

Typical sound insulation values for a range of floor construction are given in Table XI.

Timber floors will give poor sound insulation unless detailed as floating constructions (see p. 263). Where they are not well tied to the surrounding walls, and the walls are slender, flanking transmission will limit the potential of this type of floor. Where applicable, the addition of pugging must make use of a relatively dense material, e.g. dry sand free from deliquescent salts (say 80 kg/m²) and must cover the full area including narrow slots at the edges, if it is to be effective. Obviously, the lower skin of the floor must be capable of carrying the load.

Dense concrete floors give very good sound insulation. Even so, they must not be thought of as a complete answer to sound insulation problems. There will be occasions when, as a result of poor noise zoning, we may require even higher performance, which can only be achieved by separated structures, e.g. for mechanical plant rooms over bedrooms. The reduced mass of hollow planks, waffle slabs, prestressed or poststressed construction is reflected in reduced performance. Joints must be well filled and sealed to get good results. A screed topping and a plastered soffit can help with this. With dry lining to the soffit, it may be necessary to seal joints from below before its application, where good sound insulation is needed.

The effect of floor construction on vibration isolation measures have been referred to on p. 151. Impact noise reduction is discussed on p. 241.

(b) Impact Isolation at Floors

Impact-noise transmission through floors is best dealt with at source, before it spreads in the structure. This can be done by using resilient floor coverings such as thick carpet on resilient backing, ribbed rubber, sheet flooring on felt or foam or thick cork tiles.

There is some variation depending on floor construction, e.g. filled timber floors help isolation. To counter low-frequency impact (a 'thump' rather than a 'click'), a floating floor is preferable. Surface treatment need not then be so soft. In this case, the base floor should be heavy. Impact noise at low frequencies is difficult to isolate in separated light timber floors. The reader is referred to current building regulations which set standards for impact isolation.

Surface noise (i.e. in the source room) is difficult to cure where smooth floor surfaces are used, although resilient materials will help.

(c) Raised Floors

To help service connections a secondary raised floor (e.g. a timber raft or metal panels on pedestals) may be mounted over the structural floor, a technique often used in offices, telephone and computer rooms. Deep raised floors occasionally provide a flanking path under partitions, particularly if acting as an air plenum with floor grilles. Impact and vibration transmission in the floor surface can be annoying and should be checked. Unless substantial and well sealed, a raised floor is not likely to add greatly to the sound insulation of the floor.

(d) Floating Floors

Floated floor constructions can give very good impact-noise isolation and in some cases improved airborne sound insulation—if carefully engineered, heavy floating floors can give over 60 dB nominal airborne sound insulation. A number of common methods are described below (see also Table XI).

A simple wood-raft floating floor consists of boards or sheets fixed to battens to form a raft which rests on a resilient quilt laid over the structural floor slab. The battens are not fixed in any way

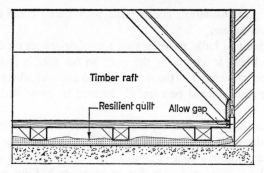

Fig. 103. Floated wood-raft floor

to the slab. Traditional construction makes use of glass-fibre or mineral-fibre quilts (about 25 kg/m³) which are selected to give a resilient bed without allowing too much movement of the floor (see Fig. 103).

A concrete screed (50 mm or more deep) can also be floated in a similar way. A waterproof layer is needed over the quilt to prevent wet concrete from running through it and the quilt is turned up at the edges. The concrete must be carefully designed and reinforced to avoid cracking or 'dishing'.

These constructions improve impact-noise isolation and may contribute a little to airborne sound insulation. Insulation of a suspended timber floor can be improved by a few dB by a floated raft (see Fig. 104). Again, the raft is laid on resilient quilt which has been draped over the joists. The battens run parallel to the main joists, sitting on or between them, so that the load is spread. Particular care is needed in levelling joists before the raft is laid. With pugging, this type of

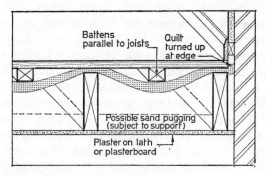

Fig. 104. Floated floor on joists

263

floor can give insulation comparable with that of 200 mm solid reinforced concrete.

It is difficult to build partitions on top of floors set on quilt. The loading is likely to compress the quilt so far that it is no longer resilient and movement of the floor tends to cause cracking. Services laid in the floor should be carefully located to avoid 'bridging' or damage.

The floors just described are relatively light and the isolation provided by the quilt is often not ideal. Where heavy floor loadings occur, or an improvement in sound insulation over that of a solid concrete floor is needed (e.g. for plant rooms), fully engineered floated concrete floors are capable of very high insulation, subject to the control of flanking transmission (see Fig. 105). Success is

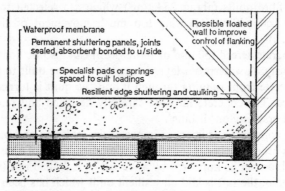

Fig. 105. Fully engineered floating floor

heavily dependant on careful selection and loading of isolation material and detailing to avoid 'bridging' and flanking. Mounts are distributed in a pattern which results in the correct loading. There are obvious advantages in using non-linear mounts, i.e. maintaining isolation over a wide range of loading. This allows for more flexibility during construction and in use (variation in loading may occur in the future, and can be accommodated within limits). Mounts and absorbent blanket infill can be bonded to permanent shuttering panels which are laid, sealed together and covered with a waterproof membrane ready to take mesh- or rod-reinforced concrete. A resilient edge trim is used to separate the floor from the surrounding structure and is cut back and top-sealed after the concrete has been poured. Plant-room floors usually need to be waterproof. Edge seals must therefore be carefully detailed to cope with this. Drains, pipes and ducts passing through floors must be detailed to avoid bridging.

Where very high sound insulation is needed, walls can be built up on the edge of the floating floor to reduce flanking. Ties to the outer wall should be avoided if possible. Vigilant supervision of this type of installation is essential. If very heavy equipment is to be brought in to a plant room over the floor, it may be necessary to spread the load on bearers until it is in position. Where sound insulation is only needed locally, e.g. immediately below a noisy fan, a small floated concrete base may be adequate. Designs must be detailed to avoid debris fouling the mounts.

Vibration isolation set on any floating floor must be selected in line with the principles referred to on p. 155.

Suspended Ceilings

Suspended ceilings can be used to improve the sound insulation of a floor construction, particularly if the performance of the floor is not already good. Where a floor is already giving mean insulation of 45 dB+, flanking transmission in walls tends to limit improvement by this method.

For such a ceiling to give useful insulation, the following features are recommended:

1. A moderate heavy membrane is needed (25 kg/m² minimum).
2. The membrane should not be too rigid.
3. The ceiling should be essentially airtight to prevent sound penetration that would occur through unsealed joints, light fittings or porous material.
4. The points of suspension should be as few as possible.
5. The deeper the air space above the ceiling, the better.

This type of ceiling tends to make access into the void awkward. Access panels must be detailed to encourage a good seal.

In spite of their value for sound absorption, many lightweight acoustic ceilings are not suitable for sound insulation because of their porous nature or inadequate mass. Plaster on expanded metal is useful provided it is of sufficient weight (e.g. 25 mm thick) and supported on light metal or spring hangers. Heavy board or plank ceilings are effective if joints are well sealed and the units are not too resonant. Light rigid boards fixed at close intervals directly on the structural slab, e.g. 10 mm plasterboard on battens, are not helpful.

I*

Table XI gives typical values for sound insulation between rooms determined by flanking via a common suspended ceiling void.

Dry Partitions

There are often advantages on building sites if wet trades can be avoided. This, in combination with a demand for demountable partitions and a preference for off-site engineering, has encouraged the development of a wide range of dry partitions.

Typically, sound insulation performance ranges from 15 or 20 dB for e.g. a single skin of plywood on a frame up to 50 dB+ for e.g. heavy laminated panels separated by, say, 100 mm, with minimum connection between the two panels (e.g. a perimeter frame only) and an absorbent quilt hung in the air space. Elaborate dry partitions can even satisfy regulations for party walls subject to the suitability of the associated structure. The performance of a number of common dry constructions is set out in Table XI. When comparing average sound insulation for different partitions, the range over which the average is taken must be checked. We should also bear in mind that laboratory tests may not be matched on site.

Most common are double-skin constructions on a frame or honeycomb core. Heavy sheet materials such as steel, plaster or asbestos may be used separately or combined to form the panels. Most light weight partitions with a honeycomb core give less than 30 dB. Performance tends to increase with panel weight, the depth of the airspace and the damping and de-coupling of the panels. The use of absorbent quilt (typically 10–100 kg/m³) is particularly helpful for lightweight partitions, usually adding 3 to 4 dB, but sometimes as much as 8 dB.

A dry construction relies on good seals at the joints and the edges. These can prove awkward where highly engineered partitions meet an uneven building shell. Naturally, there is a preference for dry sealing, where partitions are to be truly demountable. These must be compliant enough to cope with variation in the width of the gap. Sometimes special measures are needed e.g. when a partition butts up to a ribbed ceiling tile—filling the tile may be the best method in this case. Where 'wet' seals are permissible, these tend to give better results, if properly applied. Complicated metal assemblies used in some demountable partitions are prone to weakness due to small air paths and must be carefully examined.

Of course, selection and construction of dry partitions must allow for many other criteria, e.g. fire resistance.

Sliding, Folding Partitions

Sound insulation of sliding, folding partitions is often overestimated. Unfortunately, the need to move the partition affects the ability to seal it, except by means of a separate mechanism. Concertina constructions tend to be weak at hinges (unless they are covered with flexible insulating material) and flaps, or skirts, which cover gaps at top and bottom do not seal well. If large, and sufficiently heavy, panels are used, the few joints can be sealed separately, once in position. Performance over 40 dB can be achieved in this way. But panels must be compressed together across good seals and seals must also be extended from the panels at top and bottom. To avoid gaps at corners, the mechanism must be able to seal in the correct sequence. Success therefore depends a great deal on the mechanism, which must be robust. People may not make full use of it unless the partition is designed so that it has to be used. Where very high performance is needed, a double line of partitions (separated by a deep air space or a corridor) may be appropriate.

Clearly, to get the best from the partition, flanking transmission, e.g. via the ceiling, must be limited. Laboratory tests of the partitions must not be on single panels alone, but should include joints, where weakness is most likely. Site tests are preferable if flanking transmission is not significant.

Flexible Sound-Insulating Materials

Heavy, limp and flexible materials such as lead or loaded plastic can be very effective for sound insulation, and have applications where rigid material cannot be used well. They can form covers or wrapping over curved or irregular surfaces such as acoustic lagging on ducts (see p. 274). Thin sheet lead (5 kg/m^2) is useful where a barrier must be formed and sealed against irregular boundaries (e.g. in a suspended-ceiling void), or as a simple blanket overlay. Heavy curtains in dense flexible sheets can provide helpful screening around industrial processes and allow easy access. Some of these materials need to be checked with the fire authorities before use.

Seals

Seals often fail simply because they do not fill the width of the joint. Dry seals must offer good enough compliance and must be well compressed. Where compression is applied over a long length, e.g. round a door or window frame, the material needs to be more than usually soft/compliant (e.g. wipe seals or hollow sections). Foam strip materials should be of closed-pore construction. For special applications magnetic or inflatable seals may be appropriate.

'Wet' seals must adhere well to both sides of the joint and need careful application. Good seals cannot be expected where access is difficult. Adjoining surfaces must be clean, dry, and free of oil before application and a backing strip is often helpful.

The material must be durable, accommodate temperature change, retain adhesion and compliance and resist chemical attack. Currently some of the best sealants for sound insulation are based on polysulphides, silicones, urethanes and acrylics. Although often expensive, failure is often more so. Oversize holes or joints should be reduced in size before attempting to seal them.

A wide variety of grommets, gaskets and cover strips is available for holes, gaps and joints. Where pipes pass through walls, perhaps we need to take account of pipe temperature, thermal movement and possibly vibration isolation. This is an example, where, as with all seals, careful detailing is vital.

Acoustic Doors

We first consider the potential of standard panel and solid-core doors. A very high proportion of doors are very poorly sealed. As a result, they tend not to develop their full potential for sound insulation. Fig. 106 shows how much improvement can be made simply by providing good seals.

Table XI gives outline guidance on the sound insulation performance of doors.

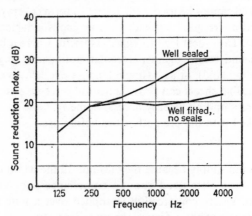

Fig. 106. Improvement in sound insulation of a simple hollow-core door when sealed

268

Door seals, even when provided, are often not of the right type or material. A sealing strip which is quite soft when a sample is pressed with the finger turns out to be very stiff when compressed over a long length by the edge of a door. Therefore, if compression seals are used, they must be very soft and so may be more easily damaged if exposed proud of the frame. Accurately aligned wipe seals which rely on a shear seal can be as good. But brush seals are not so effective. Magnetic seals (a smaller version of the type commonly used on refrigerators) are more expensive, but very effective. The most difficult parts to seal are the threshold, and, on double doors, the junction of the leaves.

There are a number of alternatives for dealing with thresholds. We can use a raised bar containing a rounded resilient strip. But to get compression on this, the door may be slightly difficult to open or close. Good alignment is absolutely vital. The threshold of a door cut close to a carpet tends to act as an acoustically lined slot and therefore gives some protection. Magnetic seals tend to pick up metallic dust and debris and may get damaged in this position. Drop bars and flaps can be used after the door is closed in special cases.

Where the leaves of double doors meet, a seal can be made on a rebate, if the doors open one way only. Double swing doors are awkward to seal at the junction. Wipe seals tend to move apart or catch on one another. Brush seals may be the only answer in these cases, even though they do not seal very well. On swing doors required to give good sound insulation, door closers should be strong enough to resist partial opening as a result of small air movements. Ironmongery must be selected to keep a good seal all round the door.

It may seen an obvious point, but sound insulation of doors can be reduced to very low values where air-transfer grilles are installed in them.

Specialist Doors

To improve insulation further, we must add mass to the door and improve the seals and the ironmongery. Timber doors can be improved e.g. by incorporating steel or lead sheet. For plant rooms and industrial use, double-skin metal doors (with absorbent packing in the cavity) may be appropriate. Small double observation windows can be set in doors with performance up to 40 dB, without significant loss of insulation. As insulation improves above 35 dB, the quality of seal becomes more important. Even the amount of compression on the seals becomes significant.

Ironmongery must therefore be set up to close tolerances. Espagnolette bolts are particularly useful for this type of door. For access panels and doors used infrequently, a screw fixing may be acceptable. For double doors needing high performance, a removable centre mullion may be used to allow better seals.

For small doors, performance above 40 dB may be best obtained using two doors either side of a sound lock lobby (see Fig. 107). Heavy single sound-insulating doors giving more than 40 dB are available, but are very expensive and present a problem where quick fire escape is needed.

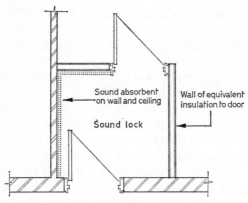

Fig. 107. A sound lock lobby

For recording studios and rear stage areas in large theatres very large, heavy acoustic doors are often necessary. These doors need power operation and often seal by using the weight of the door to compress the sealing trip by cranking it in on to the frame at the very end of its traverse.

Acoustic Screens

Screens should be designed and located on the basis of the principles discussed on p. 135.

External screens must not only be designed to ensure adequate additional path length for the sound but also be homogeneous to limit direct penetration, and be designed to cope with wind pressure and other weather exposure.

Internal screens can be of very simple construction and need not be heavy so long as they contain one continuous, impervious membrane of not less than 4 kg/m². Usually, it is helpful for surfaces to be absorbing, e.g. fabric stretched over porous absorbent material or, where impact damage is likely, with perforated-sheet facing replacing the fabric. Some open-pore foams which have been used for this purpose may produce toxic fumes in a fire and are therefore not favoured. Materials such as mineral wool or glass-fibre are preferred.

Screens should be detailed to meet walls, furniture or other screens without leaving large air gaps. If fixed to furniture close to the source or listener (e.g. desks in offices) they can be very effective. The side facing the source/or listener should be absorbent.

It is important not to place too much reliance on portable screens, which tend to get used to define territory, as vagrant pinboards, or in positions where strong bypass reflection can occur. A survey of portable-screen performance under typical conditions of use, e.g. in an open-plan office, tends to yield mean attenuation of about 5 dB at mid-frequencies.

Partial Enclosure

Much of the noise from a wide range of machinery can be limited by partial enclosure. Because of the need for ready access and a build-up of heat from machinery, it is often not possible to complete the enclosure. Partial enclosure is an extension of screening around the source and should be designed to cover the maximum possible solid angle subtended at the source, without interfering with access (see Fig. 46). Such shields should be lined internally with absorbent material to avoid local build-up of noise.

Examples of the use of this method are screens around noisy industrial or building operations (e.g. concrete breaking) or hoods around office machinery. Hoods over telephones use the same principle in two ways, by screening and local absorption of incoming noise as well as partial enclosure of the sound of the voice.

Noise Control by Absorption

The value of sound absorbents for noise control has been explained in Chapter 6. The materials available for this purpose are precisely those used for acoustic correction as described on p. 47, and listed for performance in Appendix A. Apart from acoustic performance and costs, materials must be checked for their resistance to fire, moisture, impact damage, bacterial growth, chemical attack, rot and vermin. Consideration should also be given to the means and effect of maintenance. Many porous materials lose efficiency after coating with paint which seals the surface. Even porous paints tend to fill the voids and reduce performance after a number of coats.

Materials such as mineral wool or glass-fibre satisfy most of these needs particularly when faced with perforated sheet. But it is unwise to use exposed fibrous material which might erode or harbour organic growth in e.g. hospitals or the food industry. The material may be faced with a very thin membrane to deal with this. But performance at very high frequencies is reduced.

In the majority of cases absorption is needed at middle and high frequencies, particularly in the room containing the source. But reduction in the source room over this range may not help sound insulation, where low-frequency noise is often the problem. Sometimes it is necessary to 'tailor' absorption to match noise sources which peak over a particular frequency range. Very occasionally, for pure tones, tuned cavity absorbers may be appropriate.

The absorbent may be fixed in many ways. Best results will be obtained if they are close to, and as much surrounding, the noise source as possible. Sometimes it is useful to suspend absorbents freely in the space, above head height. In this case, the effect on lighting, air movement and other services must be checked.

Background Sound Control

We have referred to the positive use of background sound to 'mask' unwelcome sound, in previous chapters. In practice, there are limits to the control possible using the sound sources normally found in buildings. The sound they produce may not relate well to the masking sound needed, or will not be tolerated by building occupants. Or arrangements to produce the appropriate masking sound may well produce a conflict with the function of the source (e.g. air noise from ventilation systems may only be useful when the volume flow is excessive).

Occasionally therefore, we are looking for alternative means of producing background sound. Perhaps this can be done using small fan units or other mechanical devices. One approach which has been used is to feed sound into a room using an array of loud-speakers fed by a sound signal generated electronically and tailored to suit the needs of the area in question, perhaps in combination with a public-address, music or fire-alarm system. But such techniques, and in particular the type of sound to be fed to the system are still subject to research.

In practice, a wide range of possibilities remains to be examined.

For example, it may be that noise in ventilation ducts can be picked up by a microphone in a noisy part of the ductwork and relayed through loudspeakers into the room (directly or via ducts), thus providing an easily controlled level of plant noise. However approached, these methods will always need to be designed and handled very carefully and in situations where this concept may be particularly relevant, the reader is advised to seek specialist advice.

Enclosures

Where practical, 'boxing-in' noisy equipment is obviously a valuable method for reducing noise. As we have said, access must be allowed for. Door (and window) performance must therefore match that of the enclosure. To allow air to be taken in and out without allowing noise to escape, some form of duct or plenum attenuation is needed. Holes for pipes and conduit must be well sealed. The value of an enclosure is very much reduced if noise levels inside it are built up and increase the required insulation, and also this local build-up of noise inside enclosures tends to feed more energy into the structure it is set on. In some cases this may be significant. It is therefore useful, and often essential, to provide absorbent lining inside enclosures (see also p. 134). When enclosures are placed very close to an existing partition e.g. in a plant room, the overall insulation is not the sum of the normal performance of the two divisions, but something less (determined by the coupling between the partitions and the position of resonances).

Normally enclosures should not touch the machinery, because this is likely to result in flanking of vibration isolation and re-radiation of sound from the panels. The panels making up the enclosure should not be highly resonant.

There are a number of enclosures available based on modular panels (many perform in the range 25–30 dB at mid-frequencies). But often, because of the variety of plant configurations, service layouts, etc., 'specials' have to be made up, making the method more expensive.

Damping

Certain forms of construction are self-damping, i.e. they resist movement in response to vibration energy. For example, a laminated construction will resist bending by setting up shear forces between the layers in its construction. The application of damping compounds to vibrating panels makes use of this principle, depending to

273

a large extent on a good bond between the materials. It helps to use materials which resist vibration transmission, e.g. visco-elastic materials. Adding mass to a vibrating panel increases the inertia and therefore reduces movement. But where short bursts of vibration are involved, the additional mass may not be helpful.

Duct Lagging

Improving the sound insulation of duct walls is not normally achieved by thermal insulation alone. We need to make use of a second separate enclosure which is not too strongly coupled to the duct—a separate-framed box, or a sound-insulating membrane over a fibrous-quilt wrapping (e.g. plaster on expanded metal or lead sheet over mineral wool or glass-fibre). Where dry materials are used, joint seals are important. Of course, if sound is retained within the duct, it may well break out beyond the lagging. If large areas need lagging it is wise to consider whether the attenuation can be achieved earlier in the system. Duct lagging is often valuable to limit noise break-out of (and break-in to) ducts in plant rooms (see Fig. 108).

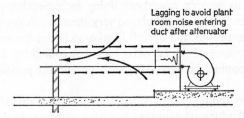

Lagging to avoid plant room noise entering duct after attenuator

Fig. 108
Acoustic lagging to limit noise entry into or exit from ducts

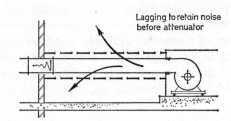

Lagging to retain noise before attenuator

Ducts and Pipes Leaving Plant Rooms

Untidy bunching of pipes and ducts as they leave plant rooms can make it extremely difficult and sometimes impractical to make a seal around them. Fortunately, a seal is often needed for other reasons, e.g. as a fire barrier. Clearly, if ducts and pipes are laid out neatly, with adequate space between them, sealing for good sound insulation becomes easier. Where, for some reason, noise is able to escape into service shafts, access panels in sensitive areas will need careful attention.

Duct Lining and Attenuators

Sound-absorbent duct lining must offer high absorption coefficients, resist erosion by the air, have a sufficiently smooth surface for low air friction and must resist moisture, fire, chemical attack and vermin. Duct attenuators tend to be of rectangular or circular section with absorbent lining and absorbing 'splitters' or 'pods' set in the airway. Rectangular attenuators with closely spaced splitters tend to give higher performance than circular attenuators of the same length, but usually with a pressure-loss penalty. Size of units is determined by the attenuation needed and limitation of air velocity to avoid large pressure loss and generation of noise in the airways. Units should be carefully located to avoid break-back of noise into the duct beyond them e.g. set in plant rooms walls (see Fig. 108). They should also be kept clear of positions where 'jetting' of air from narrow airways on to obstructions in the duct, e.g. dampers, could cause noise.

Acoustic Louvres

Louvres which incorporate absorbent material can be useful where space is limited. Usually about 300 mm deep, they can provide valuable attenuation to intake and discharge noise (typically 15 dB at mid-frequencies). They can also be combined to form attenuating screens around external mechanical plant with large air requirements, e.g. cooling towers.

The need for acoustic louvres should be identified as early as possible, because, for one reason, they must tie in with the external appearance of the building.

Vibration Isolation

Many vibration isolation systems have failed in practice as a result of simple errors. Perhaps the most common is the incorrect loading of mounts, resulting in over-compression and loss of isolation. In selecting mounts, account must be taken of the distribution of the weight of the mounted machine. To counter uneven loading or to reduce the movement of vibrating equipment, an inertia block (e.g. a solid block of concrete between machine and mounts) may be helpful. An example where this is helpful is on a fan, which, during operation, takes a back thrust in reaction to its performance. Some pre-tilting on the mounts is also possible in this case. Loading is often incorrect on sheet isolation material, e.g. cork or rubber mat, which can be too stiff over a large area of compression. This is often a fault with pump bases—the use of springs, rubber-in-shear mounts or smaller pads of material is often more appropriate. The same principle applies to a typewriter on anti-vibration mounting. The limiting factor here is usually the need to limit deflection to maintain stability.

Another obvious failure occurs when bolts, brackets, conduit or debris 'bridge' the mounts.

Flexible connections on piped services (e.g. from pumps or chillers) will not give effective isolation when they lose flexibility as a result of high internal pressures. In these cases, the problem must be tackled at source or by separating pipes from the structure. It should be noted that very large pipes will re-radiate energy as airborne noise without the help of the structure. Normally, however, we are concerned to avoid connection to the structure, which may re-radiate the energy efficiently as airborne sound. Resilient pipe clamps or spring hangers must allow for thermal movement and should be designed and adjusted after allowance for the fluid in the pipes. If not the addition of the fluid will often compress the mounts too far.

Vibrating pipes passing through partitions must have an isolating sleeve. Again, the effect of adding the fluid to the system must be considered.

Hearing Protection

In buildings, there is still a wide variety of circumstances where very high noise levels of 85–110 dBA occur and hearing protection should be worn. There are two common types which, when well fitted, give attenuation in the range 20–35 dB at mid-frequencies: (a) insert types, i.e. plugs which are inserted into the ear canal; (b) ear muffs, i.e a heavy cup around the whole ear, kept in place by headbands.

Dry cotton wool is very poor as an ear plug. Dense compressible material with slow elastic recovery is preferred. Many plugs only seal properly when they become uncomfortable. Plugs must also be discarded after use or carefully washed before re-use.

Ear muffs should be heavy and well sealed around the ear. The seal is sometimes tricky where glasses are worn. It is sometimes necessary to see that a person is wearing hearing protection. With this type there is no doubt. People cannot be expected to hear danger warnings so well when protection is worn.

Because of their great importance, whatever types of ear defenders are worn, their performance must be known.

Noise from Construction and Demolition

Construction and demolition of buildings are in themselves noisy. Particularly significant noise sources are pile driving, hammering, riveting, cartridge fixing, concrete breakers, dumper trucks, excavation equipment, compressors and concrete mixers. Increasing public concern has led to more careful monitoring of noise on sites by local authorities and more attention has been paid to the practical means of limiting such noise.

A lot can be done by careful selection of site equipment. Piles can be 'pushed' into the ground rather than hammered, and casings can be wrapped in a sound insulating 'box'. It may be possible to replace a lot of riveting work by using high tensile bolts. Drilled fixings are likely to be quieter than cartridge fixings. Concrete breakers can be fitted with very effective noise-attenuating attachments without detriment to their performance. With careful design, vehicles can also be quietened using exhaust silencers, body sealing and damping. Compressors supplying air to site tools can be substantially enclosed, leading to a useful reduction of noise.

Site planning and the consequent location of the plant on the site can make a useful contribution. Particularly noisy activities can be enclosed or partially screened. In extreme cases operatives may need to wear hearing protection close to powerful noise sources (see p. 165).

Appendix A

ABSORPTION COEFFICIENTS

The following table of random incidence absorption coefficients presents values derived from tests by recognised authorities and is divided into three groups—common building materials (1 to 22), common absorbent materials (23 to 34) and room contents (35 to 40). The frequency range given varies subject to the range used in the tests on which the data is based. In a few instances, interpolation, based on theory, has been used to complete the data.

It must be borne in mind that sound absorption is not an intrinsic property of a material alone. Factors such as thickness, method of mounting and decorative treatment will influence absorption, as will the structures in which the material is built, particularly at low frequencies.

The values quoted under the heading NRC are noise reduction coefficients which give an indication of the performance of the material as a noise-reducing treatment (see p. 136).

	Frequency Hz							
	63	125	250	500	1000	2000	4000	NRC
COMMON BUILDING MATERIALS								
1. Blockwork (clinker concrete) unplastered, unpainted	0·1	0·2	0·3	0·6	0·6	0·5	0·5	•
2. Brickwork, plain or painted	0·05	0·05	0·04	0·02	0·04	0·05	0·05	
3. Concrete, tooled stone or granolithic	0·01	0·02	0·02	0·02	0·03	0·04	0·04	
4. Cork tiles, 22 mm thick, on solid backing	—	0·05	0·1	0·2	0·55	0·6	0·55	0·38
5. Cork floor tiles—see 6								
6. Floor, solid + 6 mm cork tiles, wood blocks, linoleum, rubber or plastic flooring	0·05	0·02	0·04	0·05	0·05	0·1	0·05	•
7. Glass—4 mm	—	0·3	0·2	0·1	0·07	0·05	0·02	
—6 mm	—	0·1	0·08	0·04	0·03	0·02	0·02	
—used as a wall finish, glazed tile or polished, marble	—	0·01	0·01	0·01	0·01	0·02	0·02	

	Frequency Hz							
	63	125	250	500	1000	2000	4000	NRC

COMMON BUILDING MATERIALS—*continued*

	63	125	250	500	1000	2000	4000	NRC
8. Granolithic, see 3								
9. Hardboard over air space see 14								
10. Lath and plaster—see 13								
11. Linoleum—see 6								
12. Marble—see 7								
13. Plaster, lime or gypsum								
—on solid backing	0·05	0·03	0·03	0·02	0·03	0·04	0·05	
—on lath or plasterboard over shallow air space (< 75 mm)	0·1	0·25	0·15	0·1	0·05	0·04	0·05	
—on lath or plasterboard over deep air space	0·15	0·2	0·15	0·1	0·05	0·04	0·05	
14. Plywood or hardboard panel mounted over 25 mm air space against solid backing	—	0·3	0·2	0·15	0·1	0·1	0·05	
—as above, with porous absorbent in air space	—	0·4	0·25	0·15	0·1	0·1	0·05	
15. Rubber flooring—see 6								
16. Stone, polished—see 7								
17. Water—as in swimming pools	0·01	0·01	0·01	0·01	0·01	0·02	0·02	
18. Windows—see 7								
19. Wood-block floor—see 6								
20. Wood boards or 19 mm chipboard on joists or battens	0·1	0·15	0·2	0·1	0·1	0·05	0·05	
21. Wood-fibre board, 12 mm, mounted against solid backing								
—unperforated	0·05	0·05	0·1	0·15	0·25	0·3	9·3	
—painted	0·05	0·05	0·1	0·1	0·1	0·1	0·15	
As above, but mounted over 25 mm air space on battens against solid backing								
—unpainted	0·15	0·3	0·25	0·3	0·3	0·3	0·3	
—painted	0·15	0·3	0·2	0·15	0·1	0·1	0·15	
22. Wood-strip floor on battens	0·06	0·1	0·25	0·1	0·1	0·07	0·07	

	63	125	250	500	1000	2000	4000	NRC
				Frequency Hz				

COMMON ABSORBENT MATERIALS

	63	125	250	500	1000	2000	4000	NRC
23. Acoustic spray plaster, 12 mm								
—on solid backing	—	0·03	0·15	0·5	0·8	0·85	0·8	0·57
—on 12 mm plasterboard over 75 mm air space	—	0·3	0·3	0·55	0·75	0·8	0·8	0·6
24. Carpet								
—2 mm nylon pile bonded on plastic backing	—	0·01	0·02	0·03	0·05	0·08	0·1	0·04
—5 mm needlepunch	—	0·03	0·05	0·05	0·25	0·35	0·5	0·17
—6 mm (medium) pile bonded on closed-cell foam underlay	—	0·03	0·09	0·25	0·31	0·33	0·44	0·25
—6 mm (medium) pile bonded on open-cell foam underlay	—	0·03	0·09	0·2	0·54	0·7	0·72	0·38
—9 mm tufted pile on felt underlay	—	0·08	0·08	0·3	0·6	0·75	0·8	0·43
—25 mm brushed-hair pile bonded on hessian backing	—	0·02	0·05	0·1	0·35	0·45	0·55	0·24
25. Curtains, medium velour —straight across solid backing	—	0·05	0·1	0·15	0·2	0·25	0·3	0·17
—200% material (50% drape) over solid backing	—	0·05	0·25	0·4	0·5	0·6	0·5	0·44
26. Glass-fibre mat								
—30 kg/m³, 25 mm	—	0·1	0·25	0·5	0·65	0·9	0·9	0·57
50 mm	—	0·2	0·35	0·65	0·8	0·9	0·9	0·67
—80 kg/m³, 25 mm	—	0·1	0·3	0·55	0·65	0·75	0·8	0·56
50 mm	—	0·2	0·45	0·7	0·8	0·8	0·8	0·69
—30 kg/m³, 50 mm								
with 5% perf. facing	—	0·2	0·4	0·75	0·6	0·4	0·3	0·54
with 10% perf. facing	—	0·2	0·35	0·65	0·85	0·85	0·75	0·67
—19 mm ceiling tiles, 3·5 kg/m² faced with open-woven glass-cloth over 300 mm cavity	—	0·4	0·55	0·7	0·8	0·85	0·75	0·72
27. Metal, perforated ceiling tiles suspended with porous absorbent material laid on top*								

* Note. Where the absorbent material is bagged, there is likely to be some reduction in absorption at high frequencies.

Frequency Hz

	63	125	250	500	1000	2000	4000	NRC

COMMON ABSORBENT MATERIALS—*continued*

	63	125	250	500	1000	2000	4000	NRC
—7% perforated	—	0·4	0·6	0·8	0·8	0·7	0·5	0·72
—25% perforated	—	0·4	0·6	0·8	0·8	0·9	0·8	0·77
28. Mineral-fibre slabs*								
50 kg/m³, 25 mm	—	0·15	0·25	0·4	0·65	0·85	0·85	0·54
50 mm	—	0·3	0·4	0·75	0·85	0·9	0·9	0·72
—100 kg/m³, 25 mm	—	0·3	0·3	0·5	0·75	0·85	0·85	0·60
50 mm	—	0·4	0·6	0·8	0·9	0·95	0·9	0·81
50 kg/m³, 50 mm								
with 5% perf. facing	—	0·25	0·45	0·75	0·6	0·4	0·3	0·55
with 10% perf. facing	—	0·25	0·4	0·75	0·85	0·8	0·75	0·70
19 mm mineral-fibre ceiling board/tile	—	0·25	0·4	0·65	0·7	0·75	0·75	0·62
29. Plaster tiles, 10% perf. with porous absorbent backing infill, foil backing	—	0·45	0·7	0·8	0·8	0·65	0·45	0·74
30. Polystyrene, expanded, 25 mm thick, spaced 50 mm from solid backing	—	0·1	0·25	0·55	0·2	0·1	0·15	0·27
31. Polyurethane flexible open pore foam, 50 mm, on solid backing	—	0·15	0·4	0·8	0·8	0·8	0·7	0·7
32. Wood-fibre ceiling tile, 18 mm	—	0·15	0·4	0·55	0·7	0·8	0·7	0·61
33. Wood wool slabs, 50 mm								
—mounted solid, unplastered	—	0·1	0·2	0·45	0·8	0·6	0·75	0·51
—as above, but over 50 mm air space	—	0·15	0·45	0·75	0·6	0·8	0·75	0·65

34. SPECIAL PANELS

	63	125	250	500	1000	2000	4000	NRC
—hardboard, 3 mm, with bitumen roofing-felt stuck to back, 5 kg/m², mounted over 50 mm air space against solid backing	0·5	0·9	0·45	0·25	0·15	0·1	0·1	
—two layers bituminous felt, about 4 kg/m², mounted over 250 mm air space against solid backing	0·9	0·5	0·3	0·2	0·1	0·1	0·1	

* Note. Where the absorbent material is bagged, there is likely to be some reduction in absorbtion at high frequencies.

Frequency Hz

	63	125	250	500	1000	2000	4000	NRC
34. (*contd*) SPECIAL PANELS								
—50 mm glass fibre/ mineral wool, 80–190 kg/m³, mounted over 25 mm air space with open metal mesh covering, > 30% perf.	0·15	0·35	0·7	0·9	0·9	0·95	0·9	
ROOM CONTENTS								
35. Air x per m³	—	—	—	—	0·003	0·007	0·02	
36. Audience, in fully upholstered porous seats, per person	0·2	0·25	0·4	0·55	0·65	0·65	0·6	
—in leather- or vinyl-upholstered seats, per person	—	0·15	0·35	0·45	0·45	0·45	0·4	
—in wood, padded seats, per person	—	0·15	0·35	0·4	0·45	0·45	0·4	
37. Large audience including aisle up to 1 metre width, per m² audience area	0·25	0·4	0·6	0·8	0·9	0·9	0·8	
38. Seats, unoccupied —fully upholstered porous seats, per person	0·05	0·15	0·25	0·4	0·45	0·45	0·4	
—leather- or vinyl-upholstered seats per person	—	0·1	0·15	0·25	0·25	0·25	0·25	
—wood, padded seats per person	—	0·1	0·1	0·15	0·15	0·2	0·2	
39. Orchestral player with instrument (average)	0·2	0·35	0·8	1·1	1·5	1·2	1·1	
40. Rostrum, portable, wood, per m² of surface	0·6	0·4	0·1	—	—	—	—	

In auditoria the audience and seating will usually provide a lot of the absorption. Unfortunately, owing to variations in seat spacing, upholstery and clothing, the absorption coefficient of seating areas will vary. The coefficients given here (all rounded off to the nearest 0·05) are intended to be as representative as possible, but variations of the order of ±10% will occur from auditorium to auditorium.

In empty auditoria, the coefficient of the seating will vary so much depending on the seat spacing and upholstery that a coefficient per m² of audience is not given here. The effect of seat spacing on the absorption per unoccupied seat is to decrease the absorption per seat at the higher frequencies as the spacing gets closer, but below 500 Hz there is not much change.

Appendix B

WEIGHTS OF VARIOUS COMMON BUILDING MATERIALS

The following are conservative estimates of the surface weights of various common building materials designed for use in broad estimation of sound-insulating properties. N.B. These data should be used with reference to the Mass Law (see p. 139) only for constructions of even surface density, and must not be used where the construction is subject to obvious weakness (e.g. perforated sheets).

Item	Material	Nominal thickness (mm)	Surface weight (kg/m²)
1.	Asphalt, rock	25	55
2.	Blocks, solid, lightweight	100	50–150
3.	Boards and sheets		
	—asbestos insulation board	12	11·5
	—asbestos cement	6	9·5
	—blockboard	25	12
	—chipboard	25	15
	—hardboard, standard	3	3·5
	—mineral-fibre ceiling board	16	6
	—plasterboard	12	10
	—plasterboard plastered with skim coat to 3 mm		
	—plywood	12	7
	—wood-fibre insulation board	12	3·5
4.	Brickwork, as laid		
	—common	110	205
	—engineering	110	254
	—flue lining	110	78
5.	Concrete—gravel or crushed-stone aggregate	150	325
	—gravel or crushed-stone aggregate + 2% reinforcement	150	345
	—clinker aggregate	150	190
6.	Felt, bituminous roofing	2·5	2·5

Item	Material	Nominal thickness (mm)	Surface weight (kg/m²)
7.	Glass	6	15
8.	Linoleum	2·5	3·0
9.	Metals—aluminium	0·91 (20 swg)	2·5
	—copper	,,	8·0
	—lead	1·25 (No. 3)	14·0
	—steel	0·91 (20 swg)	7·0
	—zinc	,,	6·5
10.	Plaster—gypsum or lime	12	18
	—on wood or metal lath	—	30
11.	Rubber	3	5
12.	Slate, sawn slab	25	72
13.	Stone (average)	per 25	50
14.	Timber—seasoned softwood	,,	12
	—common hardwood	,,	17

Appendix C

SOUND REDUCTION INDICES

As explained in Chapter 6, the Sound Reduction Index of a wall or floor can only be measured when indirect transmission is negligible, as happens in the laboratory. But in the field indirect transmission can be ignored only for constructions of 40 dB mean SRI or less. There are several publications from various laboratories giving the basic sound reduction indices of many constructions. But for practical purposes, some estimate of the indices is required, including indirect transmission. This has been done in this table, but it is clear that values given will depend to some extent on the surrounding structure and cannot be directly related to the equation on p. 233. The figures given here are fairly typical of common wall, floor and window construction of areas of the order of 20 to 50 m². Since the majority of estimates involve the use of octave bands rather than one-third octave bands, the data is presented in terms of the octave bands included in the standard test range. Mean values are based on the standard 16 one-third octaves (see p. 139).

Some typical octave-band and mean Sound Reduction Indices of common walls, floors and windows (in dB)

Type of partition	Octave-band centre frequency (Hz)					Average (100–3150 Hz)
	125	250	500	1000	2000	
WALLS						
—110 mm brick, plastered to 12 mm each side	33	36	41	51	57	45
—230 mm brick, plastered to 12 mm each side	41	45	49	56	62	50
—280 mm brick, cavity wall with butterfly wire ties	38	42	51	59	63	52
—340 mm brick, plastered to 12 mm each side	43	45	51	56	62	53
—150 mm dense concrete	33	39	44	52	60	47
—75 mm lightweight concrete 110 kg/m² block, plastered to 12 mm on each side	30	35	38	42	46	39

286

Type of partition	Octave-band centre frequency (Hz)					Average (100–3150 Hz)
	125	250	500	1000	2000	
FLOORS, CONCRETE						
—125 mm reinf. concrete + 50 mm screed	35	37	42	49	58	45
—as above but screed 'floated' on glass-fibre quilt	38	43	48	54	60	49
—200 mm reinf. concrete + 50 mm screed	38	45	47	52	60	48
—300 mm reinf. concrete + 50 mm screed	41	47	50	54	61	50
FLOORS, TIMBER						
t-&-g boards or sheet decking on joists, plasterboard + skim coat below	18	25	37	40	44	35
—floated timber raft on glass-fibre quilt on joists + 50 mm sand directly on ceiling of 3-coat plaster on expanded-metal lath	37	42	47	52	60	49
WINDOWS, ALL WELL SEALED						
—4 mm single	20	21	24	27	26	23
—6 mm single	23	25	28	32	25	27
—12 mm single	28	30	33	29	32	31
—12 mm laminated	30	32	36	36	37	36
—6 mm–12 mm air gap–6 mm	21	25	29	34	29	28
—10 mm–80 mm air gap–6 mm	28	31	36	42	45	37
—6 mm–150 mm air gap–4 mm, lined reveals	22	28	36	43	47	35
—4 mm–200 mm air gap–4 mm, lined reveals	29	35	43	46	45	39

Appendix D

OCTAVE ANALYSES OF SOME COMMON NOISES

Noise	Distance (m)	Centre frequency of octave Hz									Remarks
		31·5	63	125	250	500	1000	2000	4000	8000	
Jet aircraft	—	100	112	121	123	124	123	120	117	109	Maximum values when passing overhead at about 40 m (maximum power). No mufflers.
Single-engined jet fighter	—	92	102	114	116	116	117	115	111	102	,,
Riveting on steel plate	2	78	88	96	105	106	111	109	113	110	
Suburban train over steel bridge	6	94	94	93	99	99	95	84	81	73	
Weaving shed	—	79	83	85	87	90	90	89	88	86	Reverberant sound.
Circular saw	—	62	69	71	72	78	78	86	88	87	,,
Traffic on motorway	4 (from edge)	86	88	87	82	78	77	76	74	70	
Kerbside, main road	—	75	78	79	73	72	72	70	65	60	
Inter-city electric train	30	68	68	63	61	67	71	72	64	52	150 km/hour.
Canteen (hard ceiling)	—	48	52	54	59	67	67	61	55	49	Reverberant sound.
Male speech	1	46	52	55	59	66	65	60	52	40	,,
Typing office with acoustic ceiling	—	62	68	64	60	56	55	55	53	40	Reverberant sound. Ten typewriters.

Note. The above figures are typical only; obviously all the noise sources listed vary a lot, one to another.

Appendix E

DECIBEL TABLE

Intensity Ratio	Pressure Ratio	$\overrightarrow{\leftarrow}dB^{+}_{\rightarrow}$	Pressure Ratio	Intensity Ratio
1	1	0	1	1
0·80	0·89	1	1·1	1·3
0·63	0·79	2	1·3	1·6
0·50	0·71	3	1·4	2
0·40	0·63	4	1·6	2·5
0·32	0·56	5	1·8	3·2
0·25	0·50	6	2·0	4·0
0·20	0·45	7	2·2	5·0
0·16	0·40	8	2·5	6·3
0·13	0·35	9	2·8	7·9
0·10	0·32	10	3·2	10
0·08	0·28	11	3·5	13
0·063	0·25	12	4·0	16
0·050	0·22	13	4·5	20
0·040	0·20	14	5·0	25
0·032	0·18	15	5·6	32
0·025	0·16	16	6·3	40
0·020	0·14	17	7·1	50
0·016	0·13	18	7·9	63
0·013	0·11	19	8·9	79
0·010	0·10	20	10	100
10^{-1}	$3·2 \times 10^{-1}$	10	3·2	10
10^{-2}	10^{-1}	20	10	10^2
10^{-3}	$3·2 \times 10^{-2}$	30	$3·2 \times 10$	10^3
10^{-4}	10^{-2}	40	10^2	10^4
10^{-5}	$3·2 \times 10^{-3}$	50	$3·2 \times 10^2$	10^5
10^{-6}	10^{-3}	60	10^3	10^6
10^{-7}	$3·2 \times 10^{-4}$	70	$3·2 \times 10^3$	10^7
10^{-8}	10^{-4}	80	10^4	10^8
10^{-9}	$3·2 \times 10^{-5}$	90	$3·2 \times 10^4$	10^9
10^{-10}	10^{-5}	100	10^5	10^{10}

Example: To find the dB corresponding to a pressure ratio of 25.

$$\text{Ratio of } 25 = 2 \cdot 5 \times 10$$
$$\text{In dB} = +\,8 + 20\,\text{dB}$$
$$= +\,\textbf{28 dB.}$$

Example: To find the intensity ratio corresponding to a dB ratio of −43.

$$-43\,\text{dB} = -40\,\text{dB} - 3\,\text{dB}$$
$$= 10^{-4} \times 0 \cdot 50$$
$$= \textbf{5} \times \textbf{10}^{-5}.$$

Appendix F

LOGARITHMS (to the base 10)

Number	Log	Number	Log	Number	Log	Number	Log
10	1·00	33	52	56	75	79	90
11	04	34	53	57	76	80	90
12	08	35	54	58	76	81	91
13	11	36	56	59	77	82	92
14	15	37	57	60	78	83	92
15	18	38	58	61	79	84	92
16	20	39	59	62	79	85	93
17	23	40	60	63	80	86	93
18	26	41	61	64	81	87	94
19	28	42	62	65	81	88	94
20	30	43	64	66	82	89	95
21	32	44	64	67	83	90	95
22	34	45	65	68	83	91	96
23	36	46	66	69	84	92	96
24	38	47	67	70	85	93	97
25	40	48	68	71	85	94	97
26	42	49	69	72	86	95	98
27	43	50	70	73	86	96	98
28	45	51	71	74	87	97	99
29	46	52	72	75	88	98	99
30	48	53	72	76	88	99	2·00
31	49	54	73	77	89		
32	51	55	74	78	89		

Index